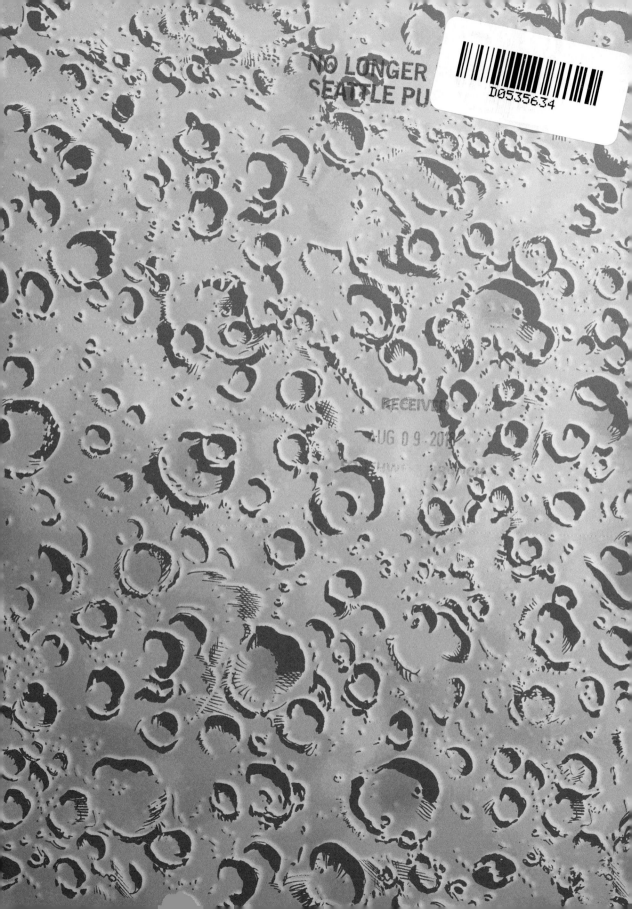

APOLLO

To our daughters,
Great journeys start with one small step.
– Matt, Chris and Mike

First published in English in 2018
by SelfMadeHero
139–141 Pancras Road
London NW1 1UN
www.selfmadehero.com

Copyright © 2018 SelfMadeHero
Written by: Matt Fitch & Chris Baker
Illustrated by: Mike Collins
Letters: Ian Sharman
Colours: Kris Carter & Jason Cardy
Appendix illustrations and design: J Francis Totti
Front cover design and typography: Dan Forde

Publishing Director: Emma Hayley
Sales & Marketing Manager: Sam Humphrey
Editorial & Production Manager: Guillaume Rater
Layout designer: Txabi Jones
UK Publicist: Paul Smith
US Publicist: Maya Bradford
With thanks to: Dan Lockwood

A CIP record for this book is available from the British Library

ISBN: 978-1-910593-50-9

10 9 8 7 6 5 4 3 2 1

Printed and bound in Slovenia

FITCH BAKER COLLINS

APOLLO

SELF MADE HERO

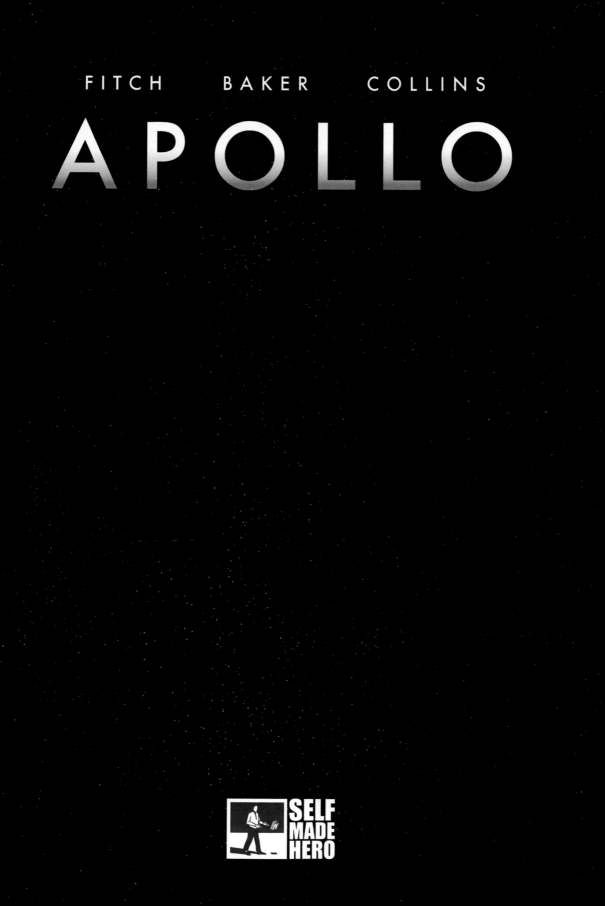

Once upon a time, we soared into the solar system. For a few years. Then we hurried back. Why? What happened? What was APOLLO really about?

- Carl Sagan

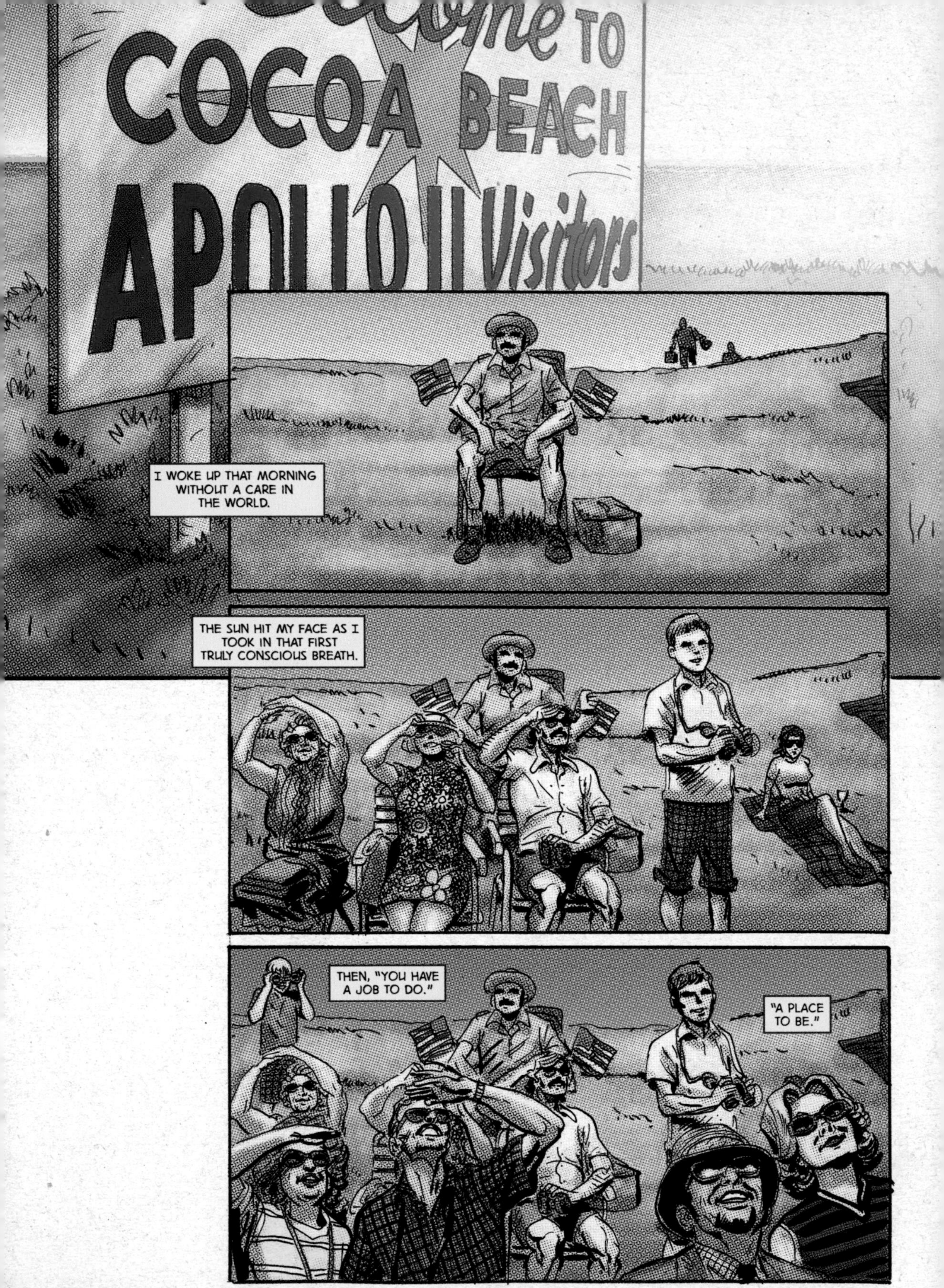
TO
COCOA BEACH
Visitors
I WOKE UP THAT MORNING WITHOUT A CARE IN THE WORLD.
THE SUN HIT MY FACE AS I TOOK IN THAT FIRST TRULY CONSCIOUS BREATH.
THEN, "YOU HAVE A JOB TO DO."
"A PLACE TO BE."

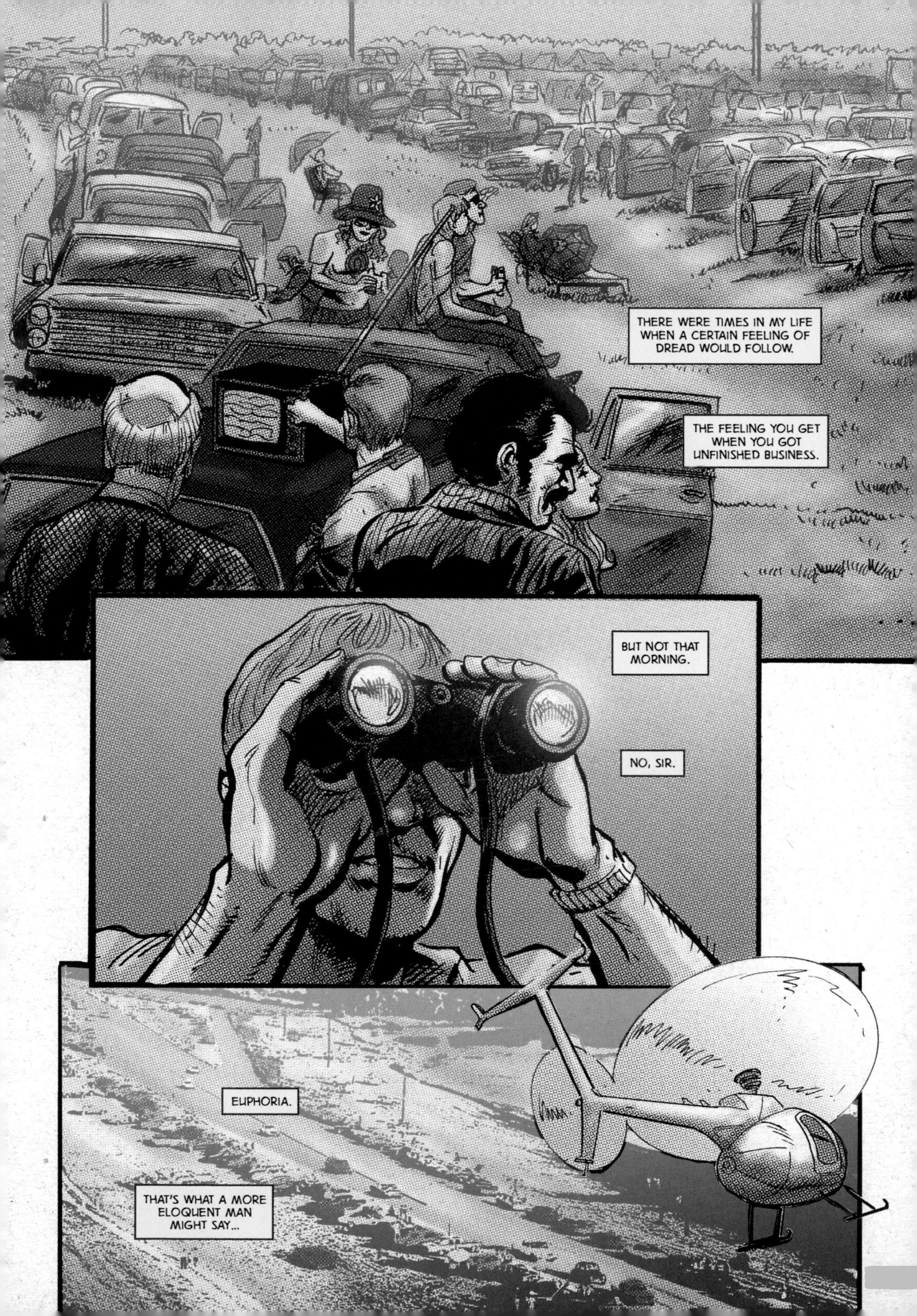
THERE WERE TIMES IN MY LIFE WHEN A CERTAIN FEELING OF DREAD WOULD FOLLOW.
THE FEELING YOU GET WHEN YOU GOT UNFINISHED BUSINESS.
BUT NOT THAT MORNING.
NO, SIR.
EUPHORIA.
THAT'S WHAT A MORE ELOQUENT MAN MIGHT SAY...

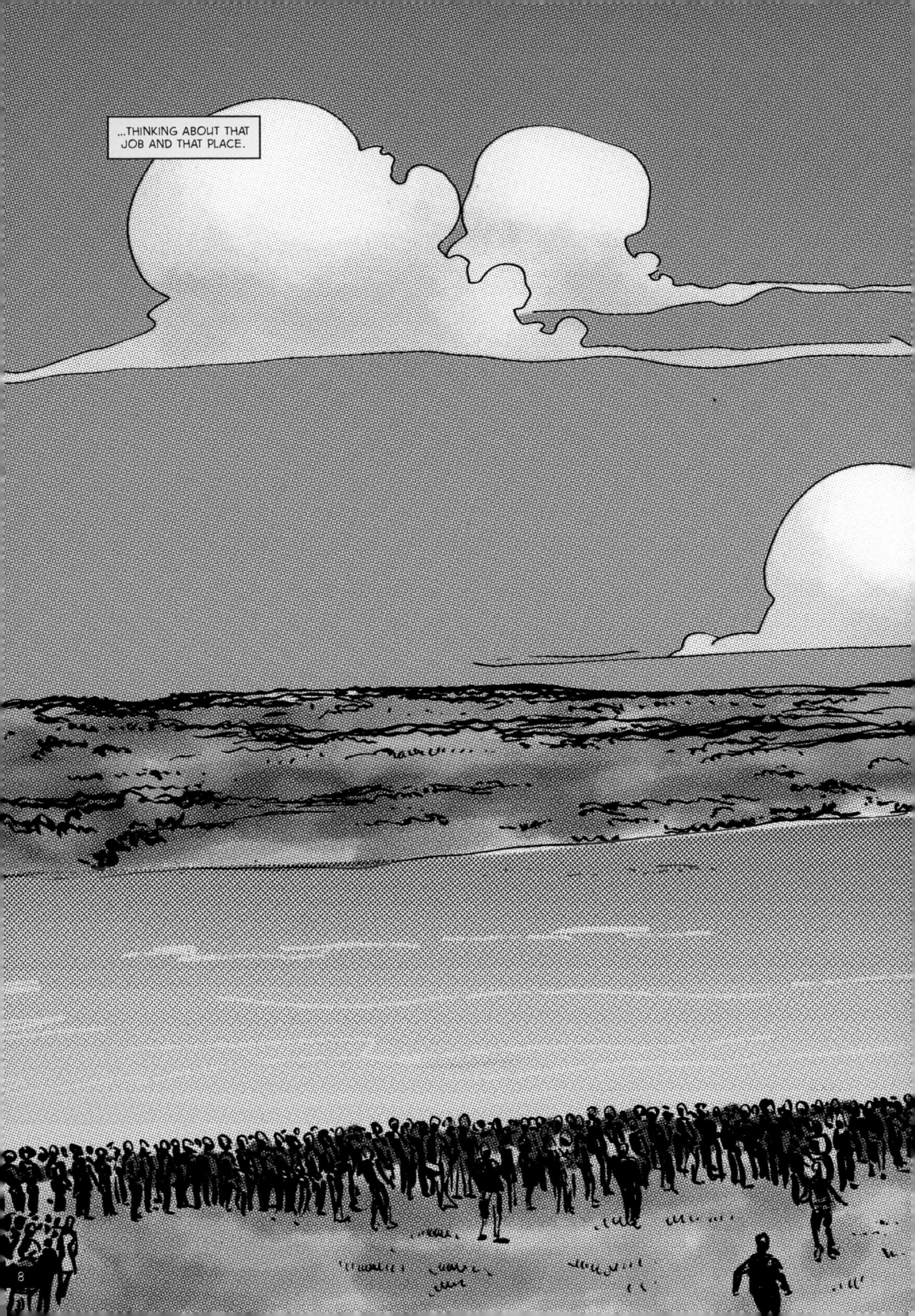
...THINKING ABOUT THAT JOB AND THAT PLACE.

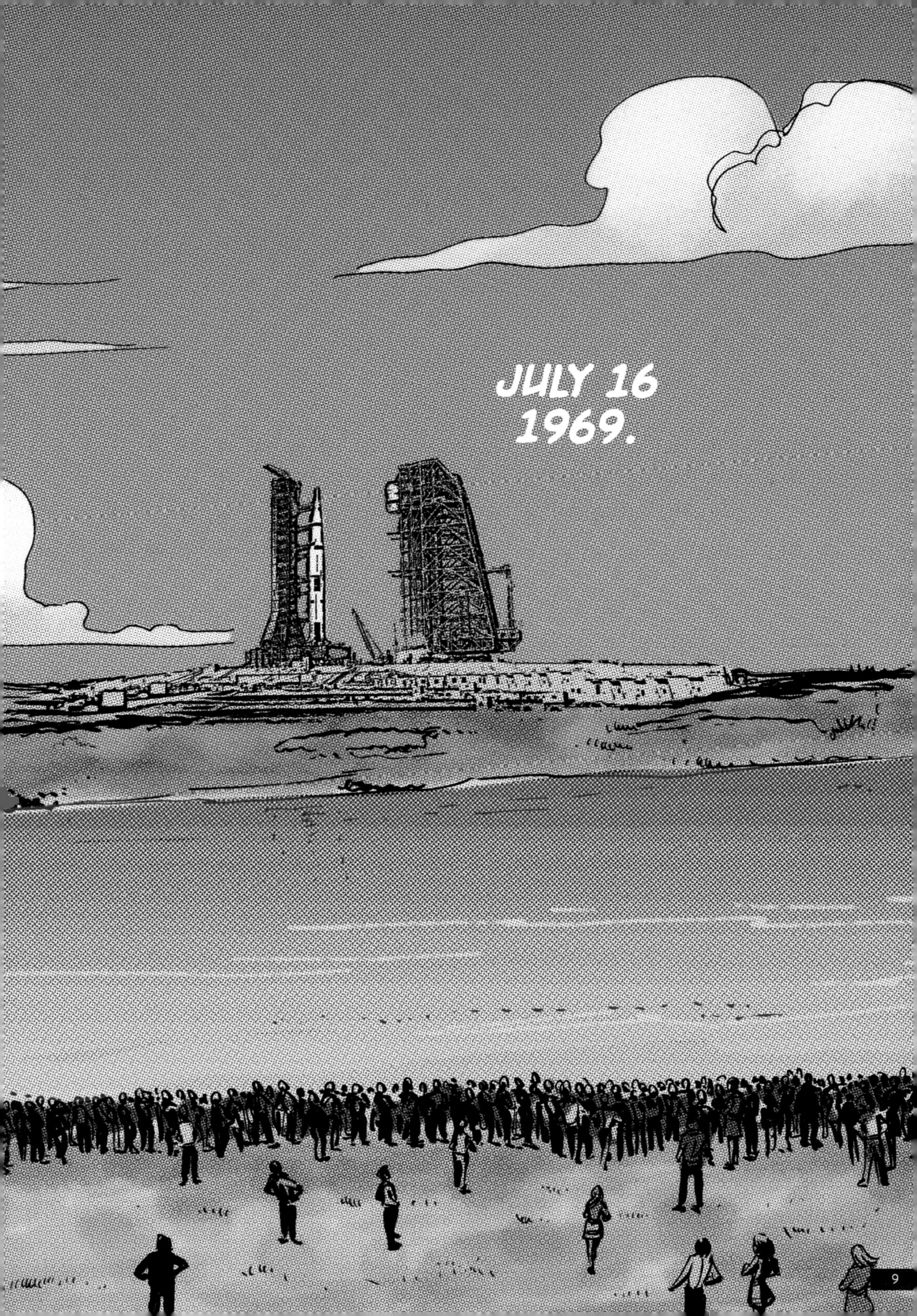
JULY 16
1969.

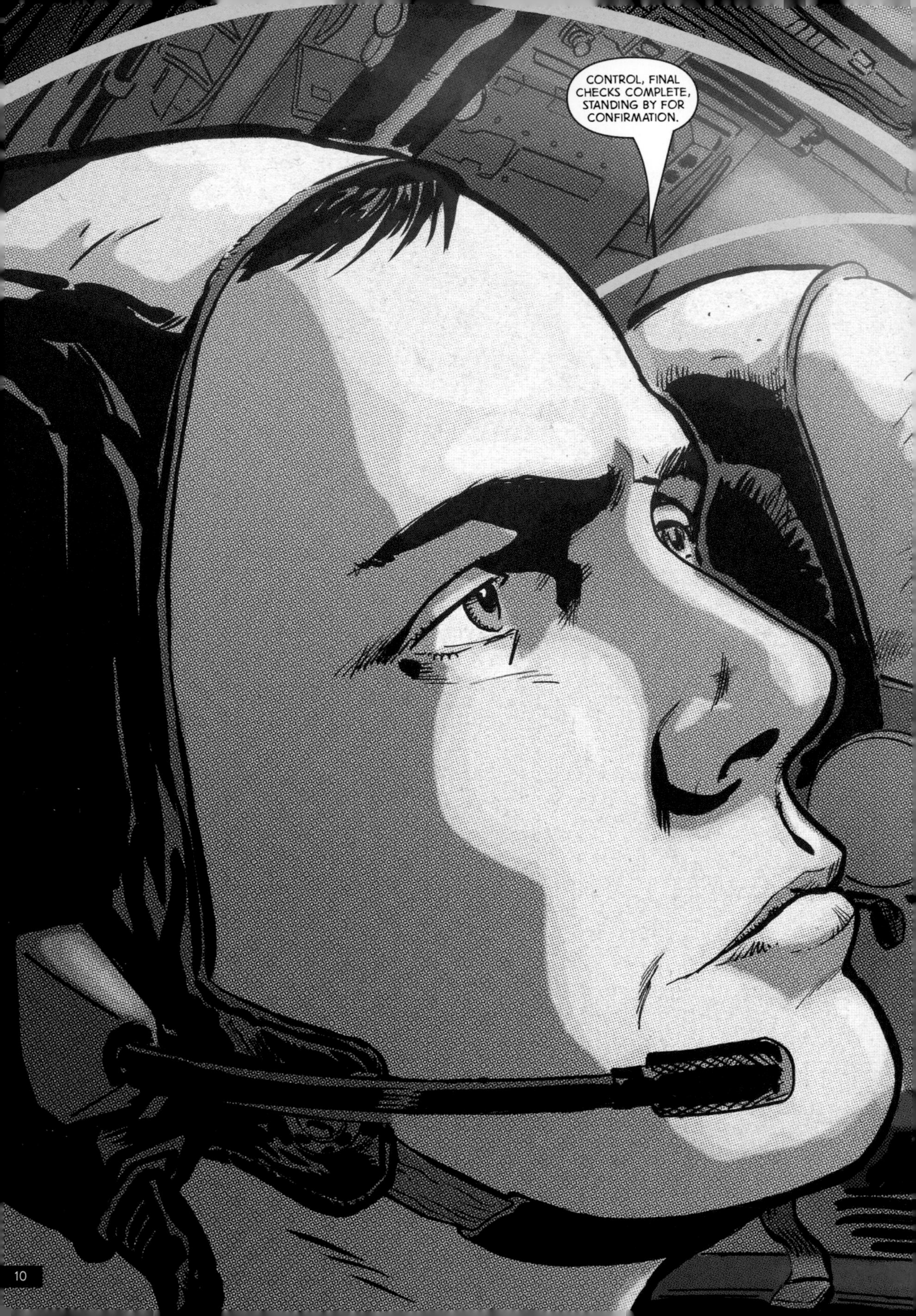
CONTROL, FINAL CHECKS COMPLETE, STANDING BY FOR CONFIRMATION.

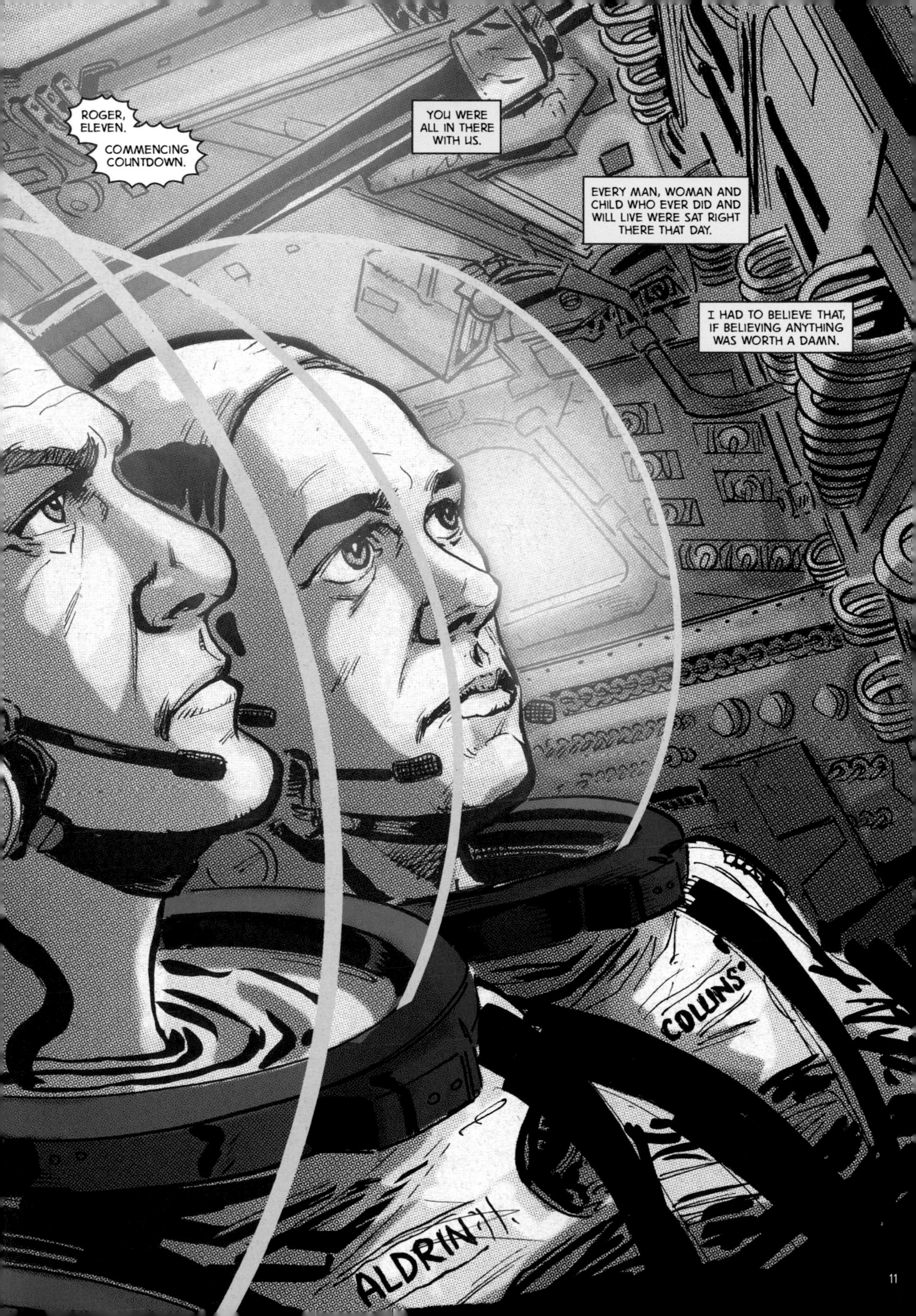
ROGER, ELEVEN.
COMMENCING COUNTDOWN.
YOU WERE ALL IN THERE WITH US.
EVERY MAN, WOMAN AND CHILD WHO EVER DID AND WILL LIVE WERE SAT RIGHT THERE THAT DAY.
I HAD TO BELIEVE THAT, IF BELIEVING ANYTHING WAS WORTH A DAMN.
COLLINS
ALDRIN

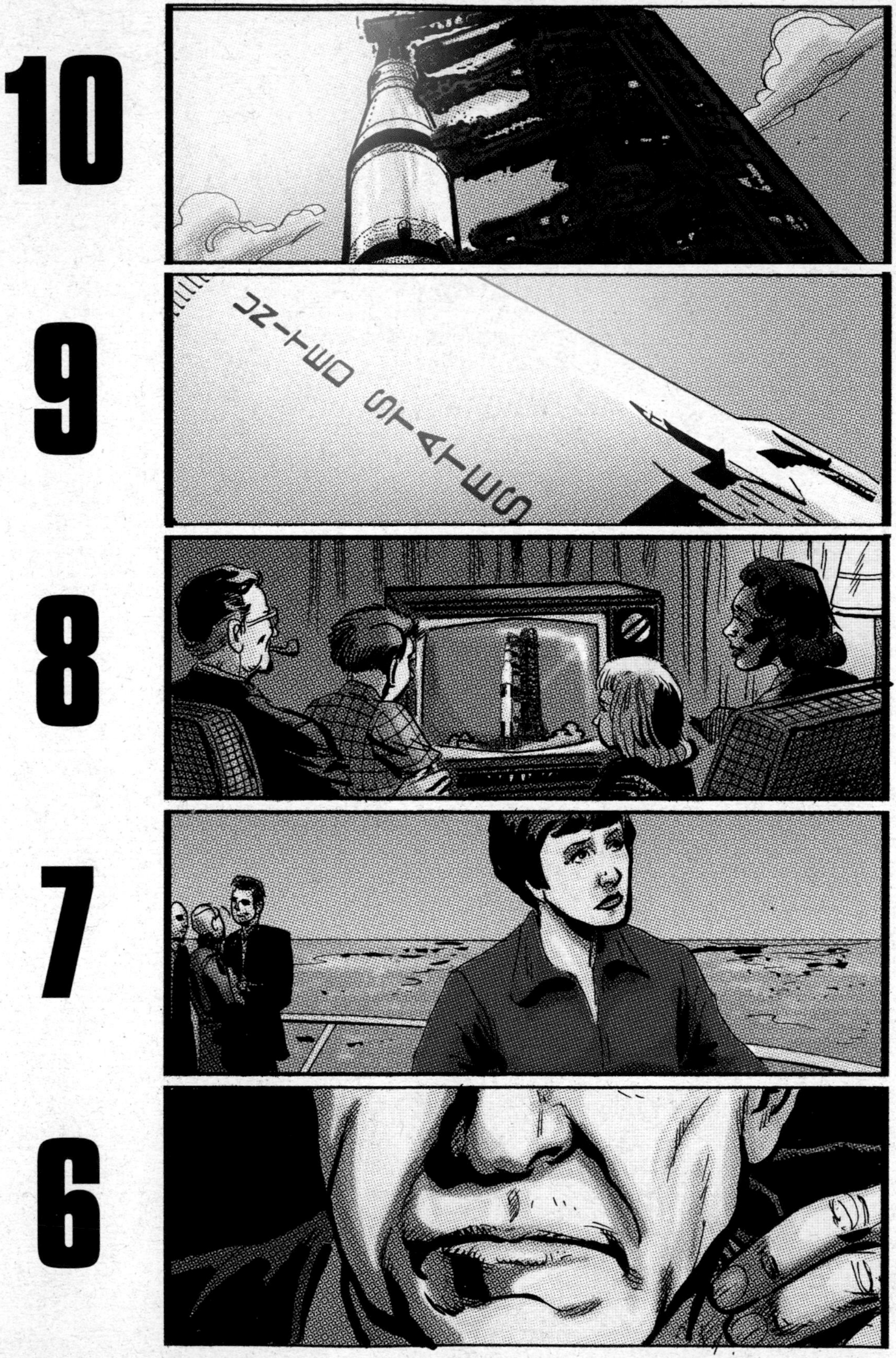
10
9
UNITED STATES
8
7
6

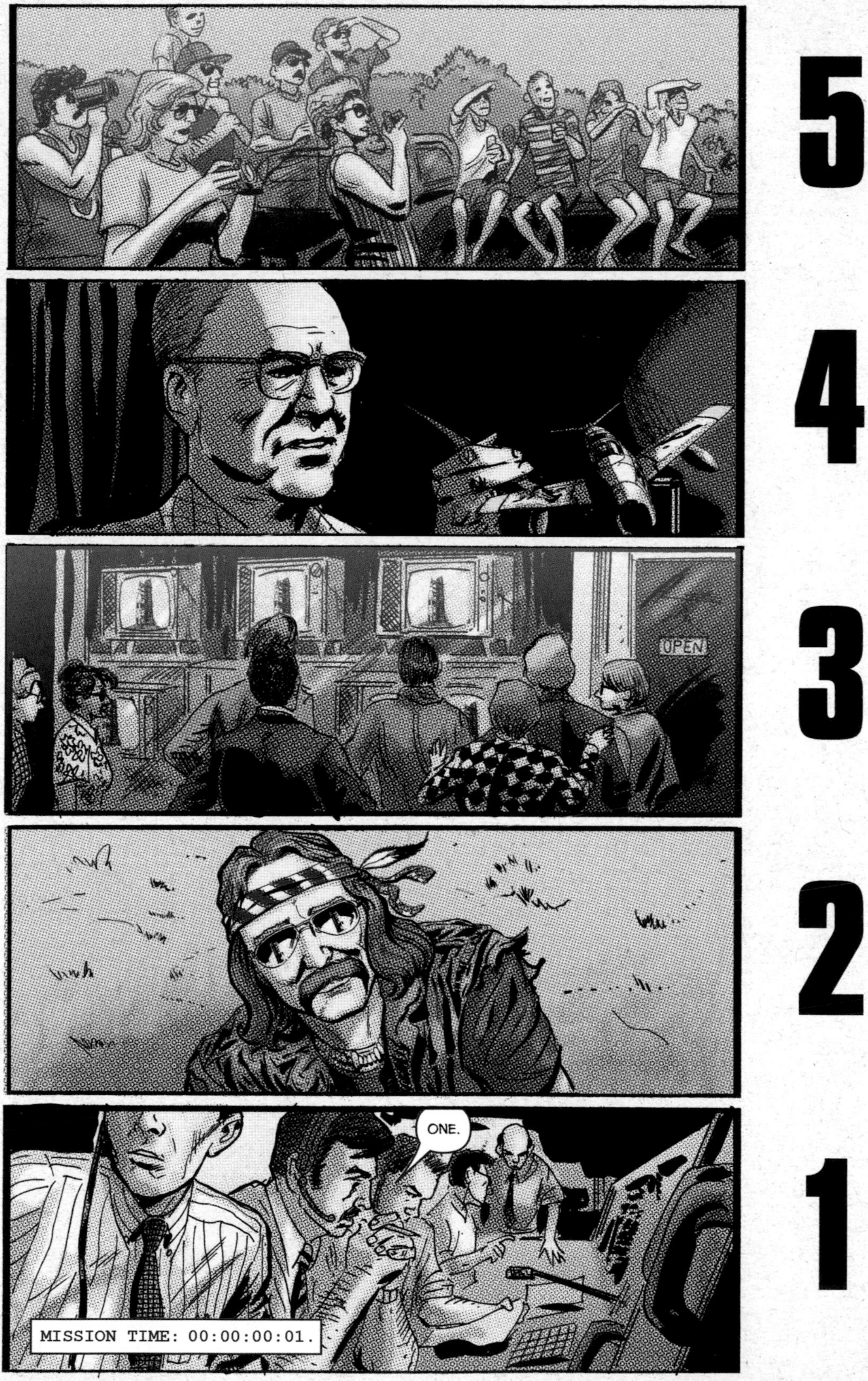
5
4
OPEN
3
2
ONE.
MISSION TIME: 00:00:00:01.
1

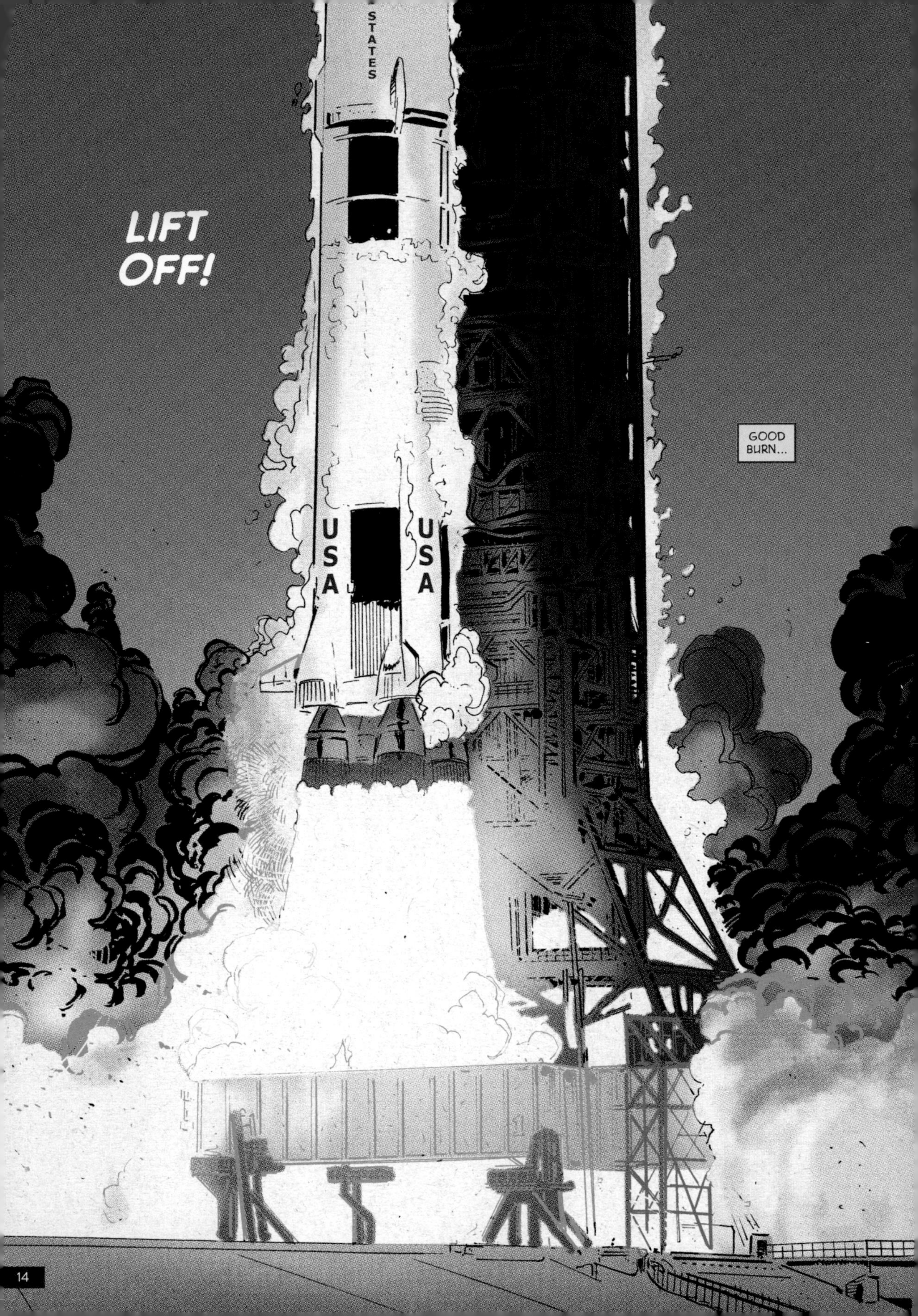
STATES
LIFT OFF!
USA
USA
GOOD BURN...

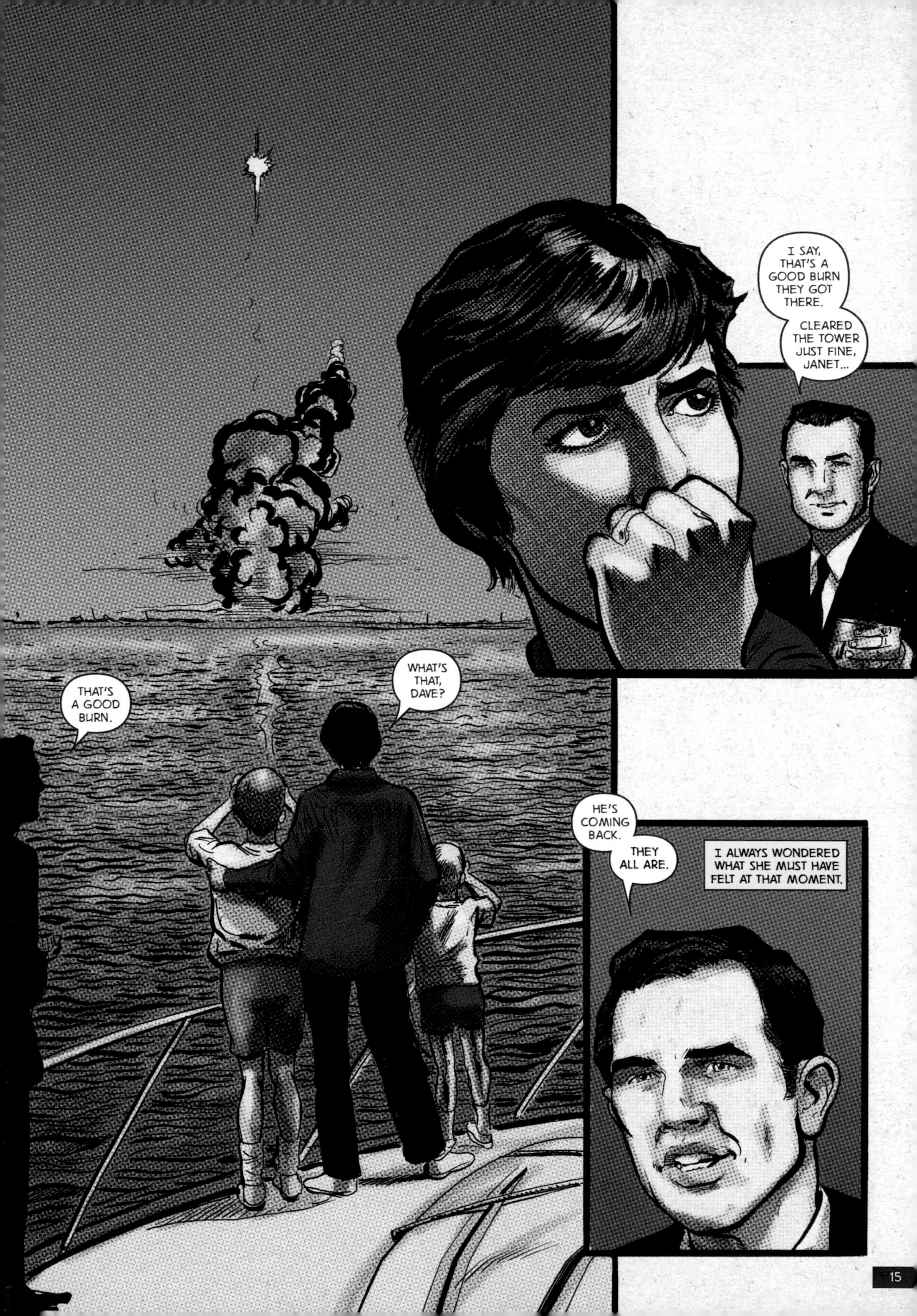

THAT'S A GOOD BURN.
WHAT'S THAT, DAVE?
I SAY, THAT'S A GOOD BURN THEY GOT THERE.
CLEARED THE TOWER JUST FINE, JANET...
HE'S COMING BACK.
THEY ALL ARE.
I ALWAYS WONDERED WHAT SHE MUST HAVE FELT AT THAT MOMENT.

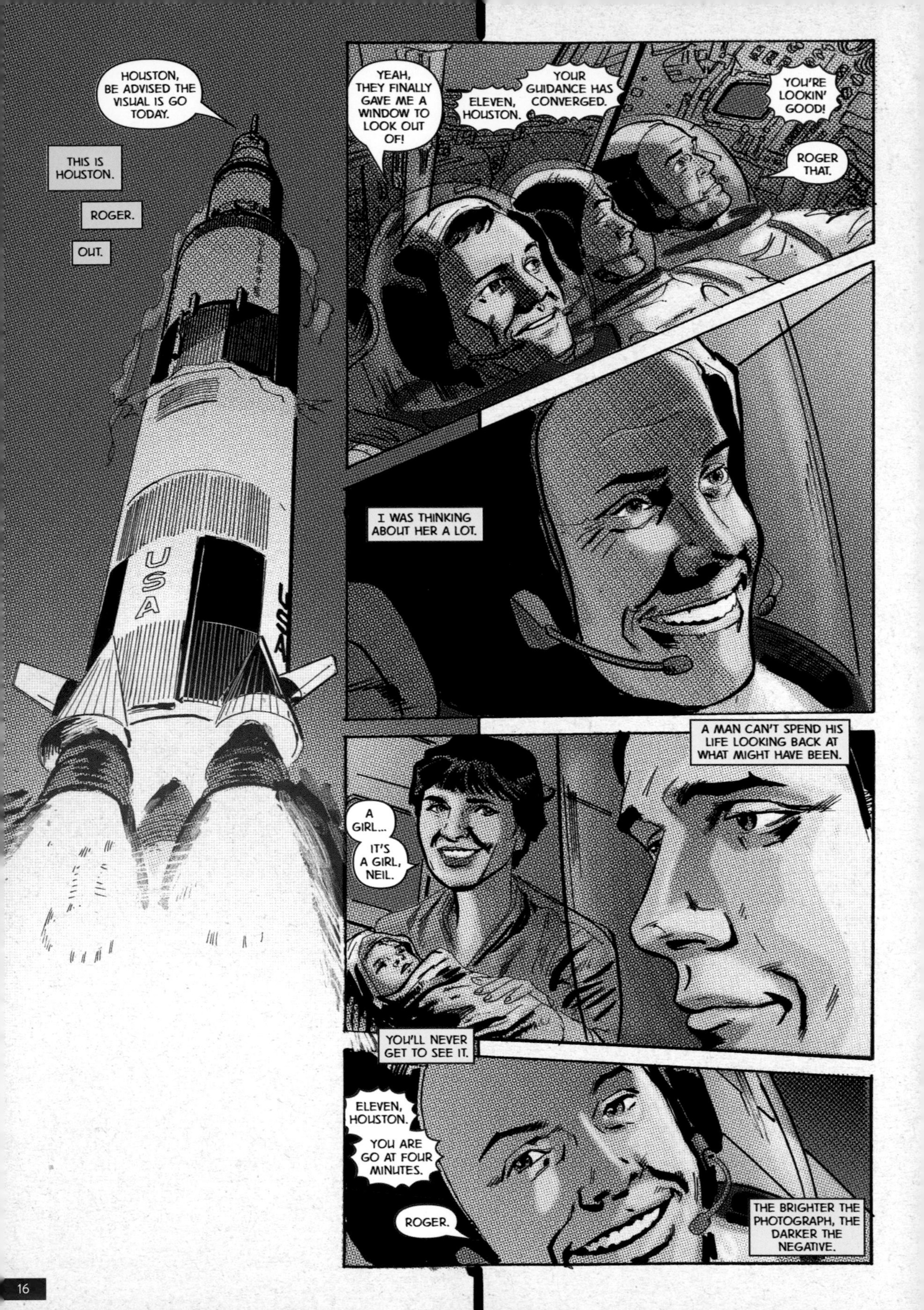
HOUSTON, BE ADVISED THE VISUAL IS GO TODAY.
THIS IS HOUSTON.
ROGER.
OUT.
YEAH, THEY FINALLY GAVE ME A WINDOW TO LOOK OUT OF!
ELEVEN, HOUSTON.
YOUR GUIDANCE HAS CONVERGED.
YOU'RE LOOKIN' GOOD!
ROGER THAT.
I WAS THINKING ABOUT HER A LOT.
A MAN CAN'T SPEND HIS LIFE LOOKING BACK AT WHAT MIGHT HAVE BEEN.
A GIRL...
IT'S A GIRL, NEIL.
YOU'LL NEVER GET TO SEE IT.
ELEVEN, HOUSTON.
YOU ARE GO AT FOUR MINUTES.
ROGER.
THE BRIGHTER THE PHOTOGRAPH, THE DARKER THE NEGATIVE.

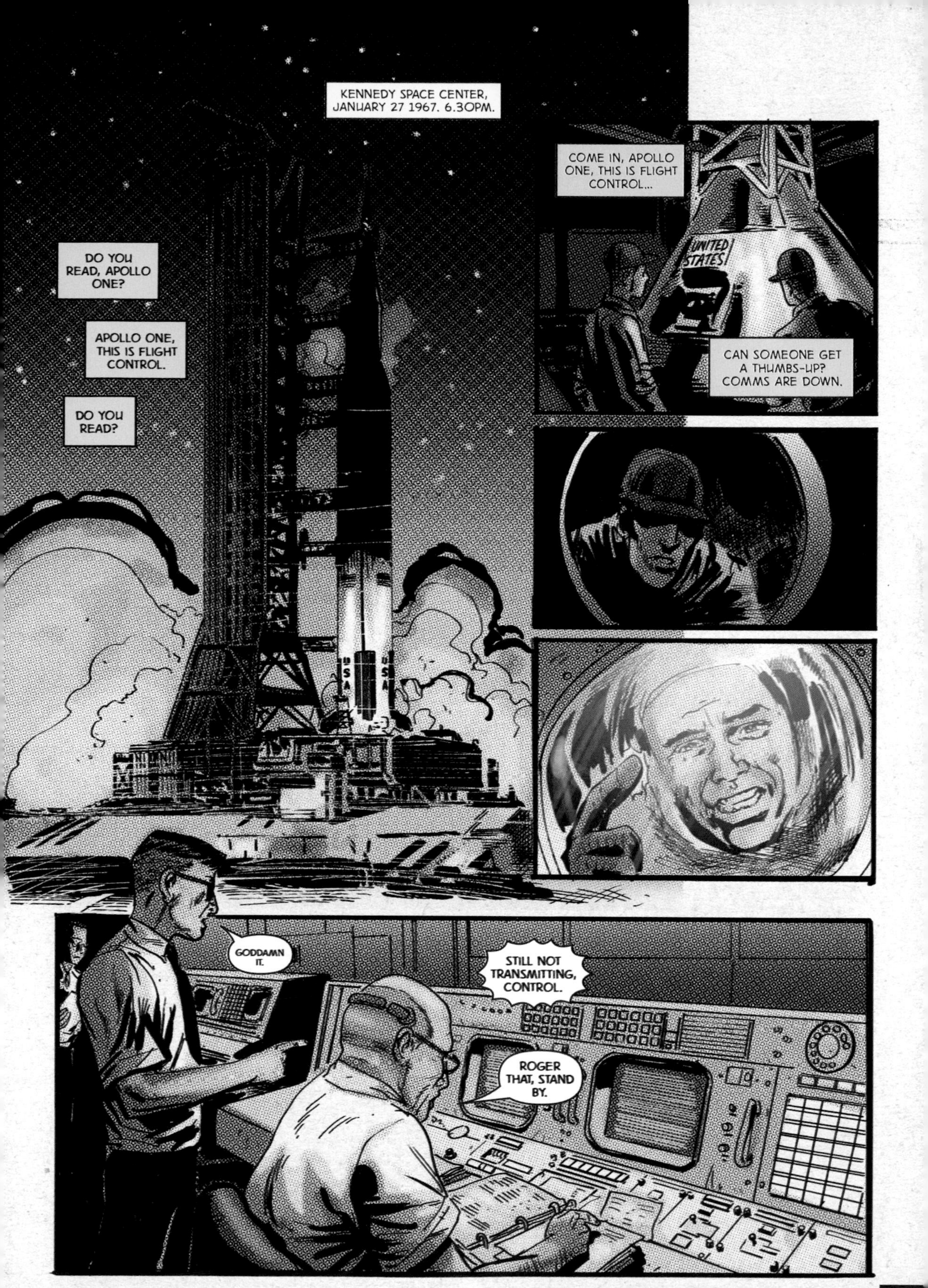
KENNEDY SPACE CENTER, JANUARY 27 1967. 6.30PM.
DO YOU READ, APOLLO ONE?
APOLLO ONE, THIS IS FLIGHT CONTROL.
DO YOU READ?
COME IN, APOLLO ONE, THIS IS FLIGHT CONTROL...
UNITED STATES
CAN SOMEONE GET A THUMBS-UP? COMMS ARE DOWN.
USA
GODDAMN IT.
STILL NOT TRANSMITTING, CONTROL.
ROGER THAT, STAND BY.

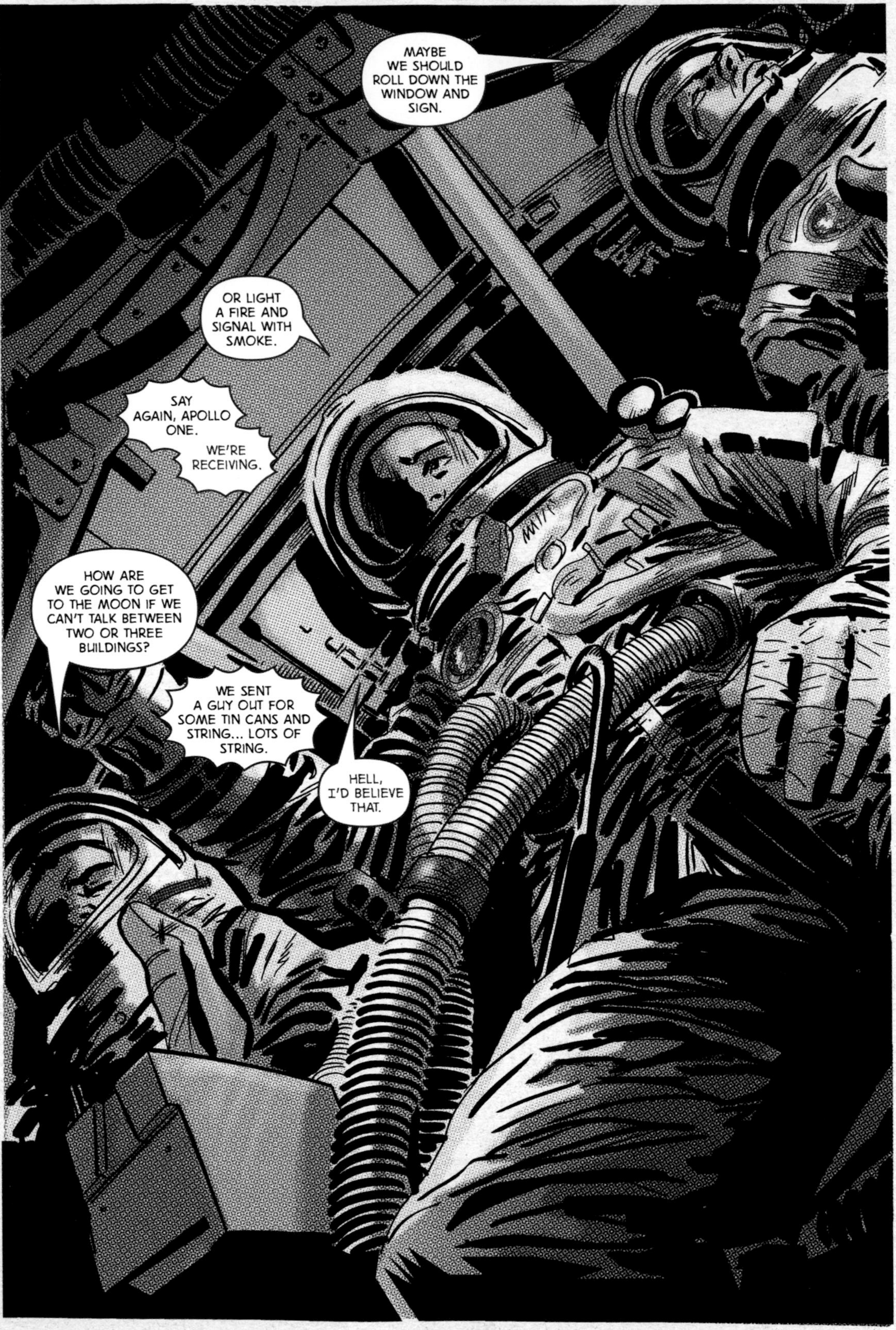
MAYBE WE SHOULD ROLL DOWN THE WINDOW AND SIGN.
OR LIGHT A FIRE AND SIGNAL WITH SMOKE.
SAY AGAIN, APOLLO ONE.
WE'RE RECEIVING.
HOW ARE WE GOING TO GET TO THE MOON IF WE CAN'T TALK BETWEEN TWO OR THREE BUILDINGS?
WE SENT A GUY OUT FOR SOME TIN CANS AND STRING... LOTS OF STRING.
HELL, I'D BELIEVE THAT.

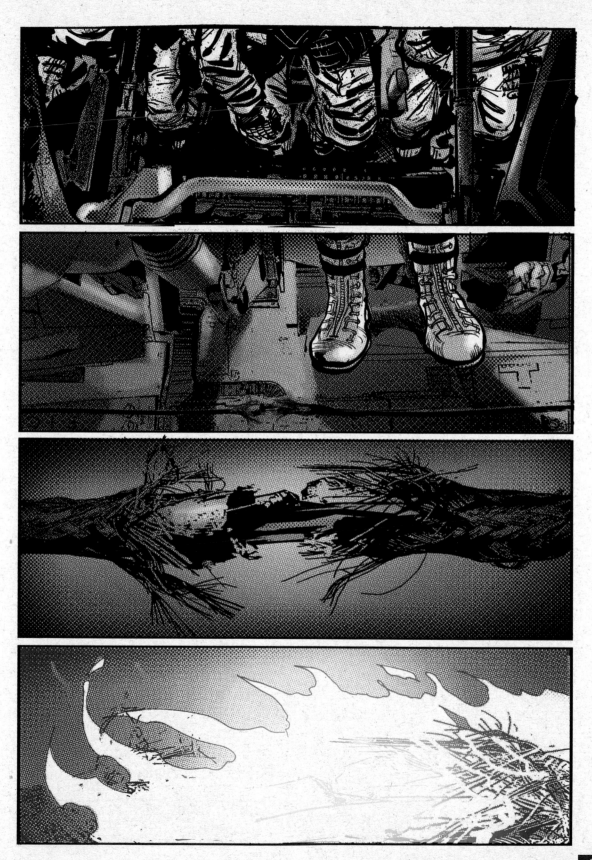

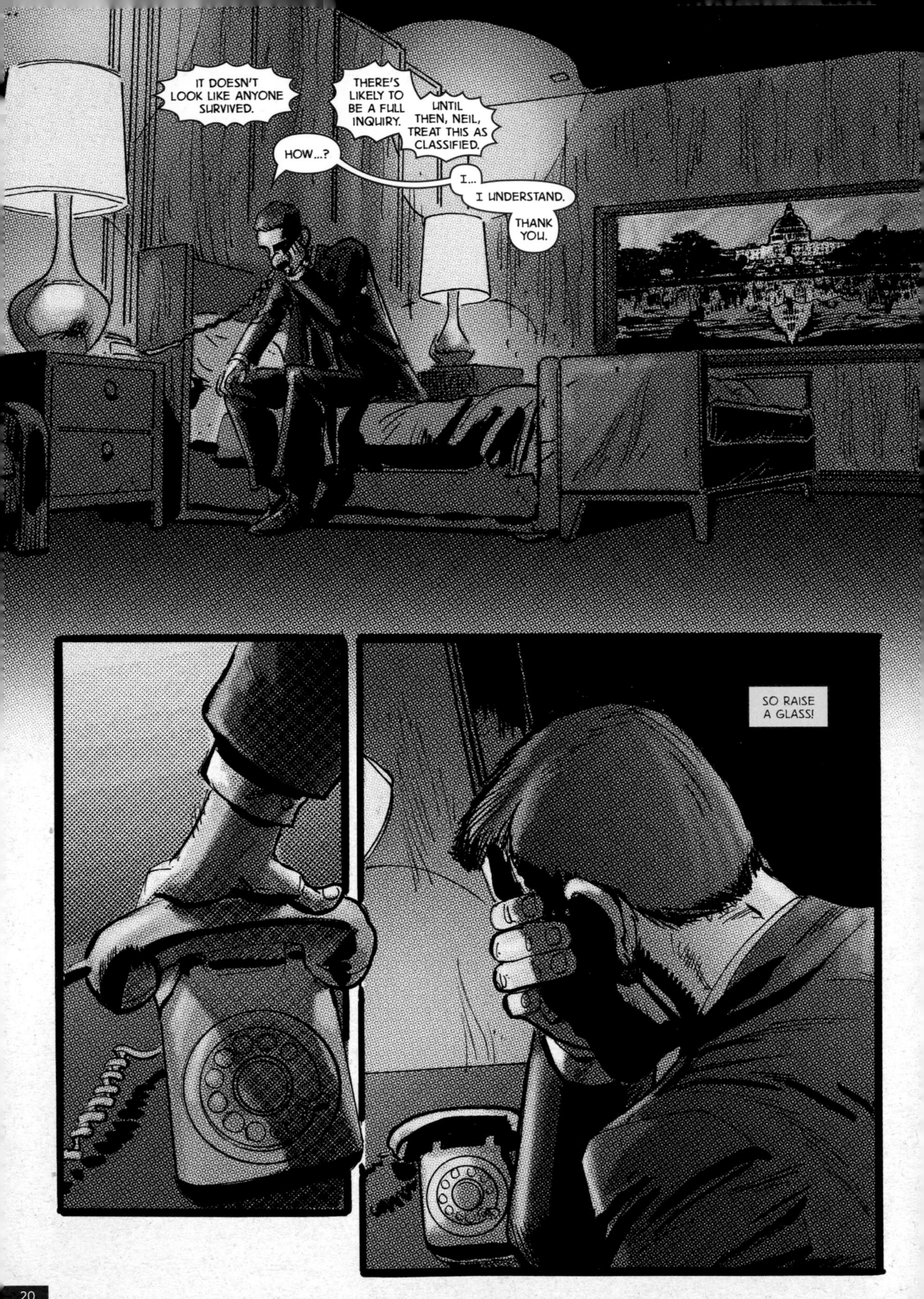
IT DOESN'T LOOK LIKE ANYONE SURVIVED.
HOW...?
THERE'S LIKELY TO BE A FULL INQUIRY.
UNTIL THEN, NEIL, TREAT THIS AS CLASSIFIED.
I...
I UNDERSTAND.
THANK YOU.
SO RAISE A GLASS!

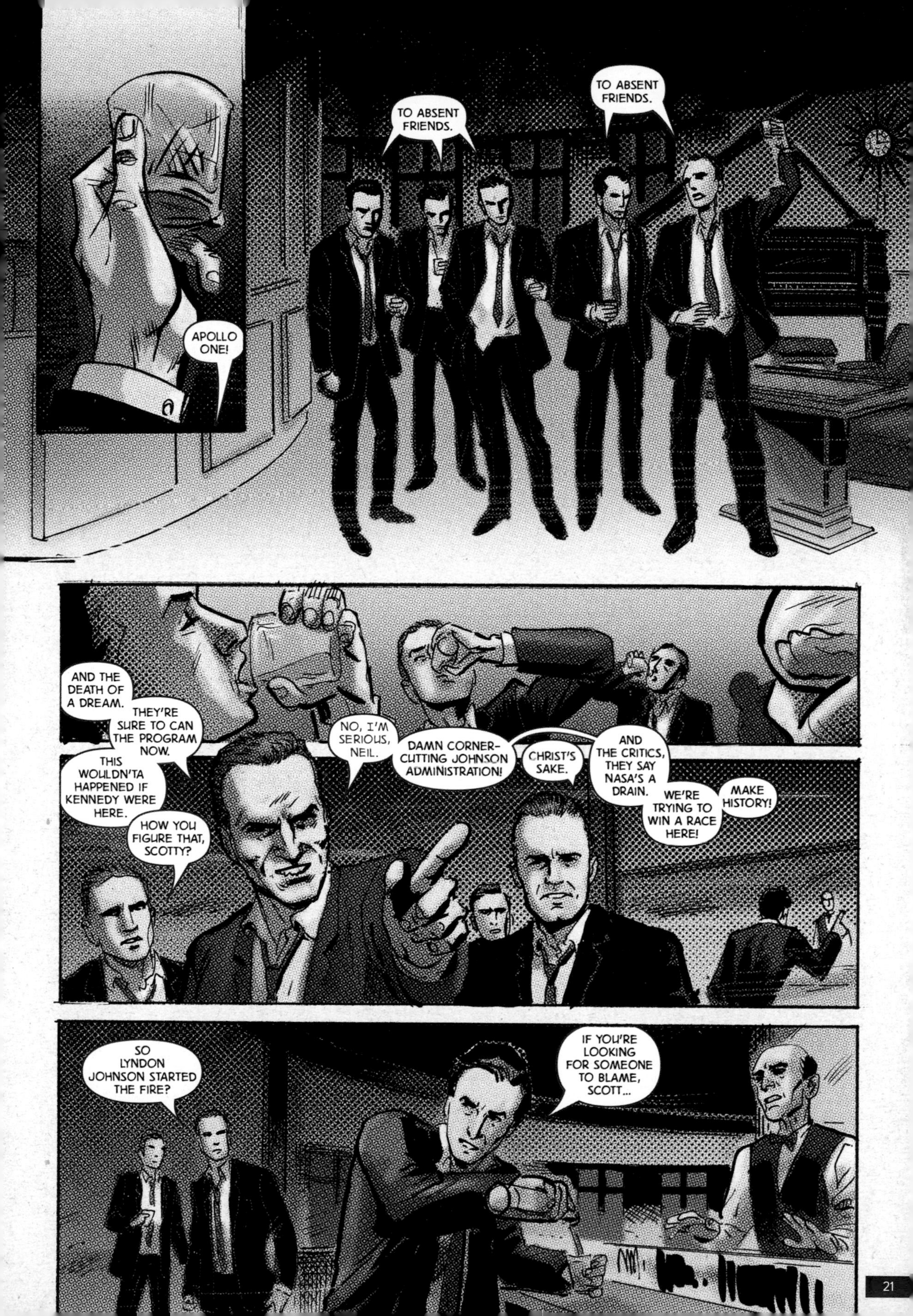

APOLLO ONE!
TO ABSENT FRIENDS.
TO ABSENT FRIENDS.
AND THE DEATH OF A DREAM.
THEY'RE SURE TO CAN THE PROGRAM NOW.
THIS WOULDN'TA HAPPENED IF KENNEDY WERE HERE.
HOW YOU FIGURE THAT, SCOTTY?
NO, I'M SERIOUS, NEIL.
DAMN CORNER-CUTTING JOHNSON ADMINISTRATION!
CHRIST'S SAKE.
AND THE CRITICS, THEY SAY NASA'S A DRAIN.
WE'RE TRYING TO WIN A RACE HERE!
MAKE HISTORY!
SO LYNDON JOHNSON STARTED THE FIRE?
IF YOU'RE LOOKING FOR SOMEONE TO BLAME, SCOTT...

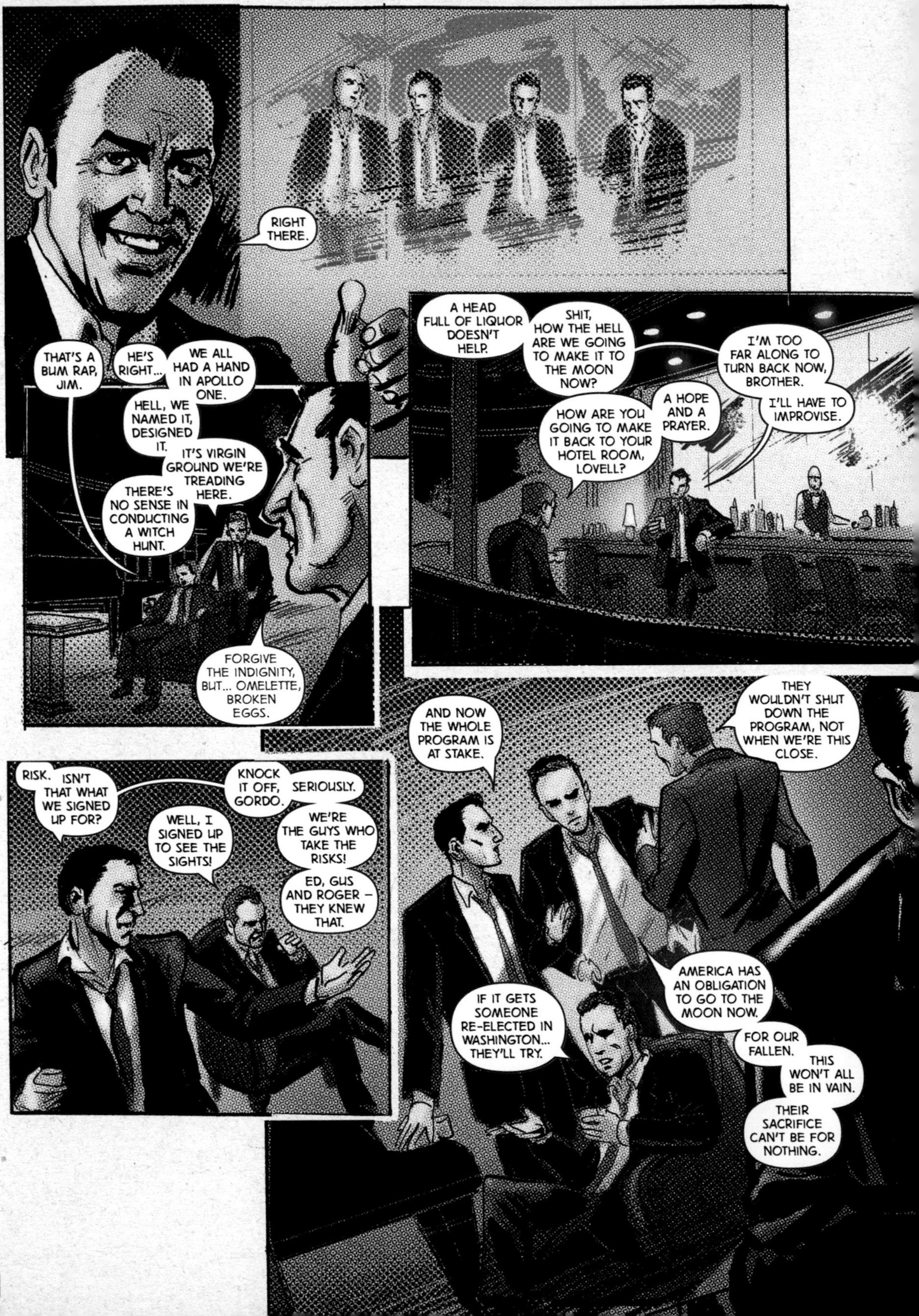
RIGHT THERE.
THAT'S A BUM RAP, JIM.
HE'S RIGHT...
WE ALL HAD A HAND IN APOLLO ONE.
HELL, WE NAMED IT, DESIGNED IT.
IT'S VIRGIN GROUND WE'RE TREADING HERE.
THERE'S NO SENSE IN CONDUCTING A WITCH HUNT.
FORGIVE THE INDIGNITY, BUT... OMELETTE, BROKEN EGGS.
A HEAD FULL OF LIQUOR DOESN'T HELP.
SHIT, HOW THE HELL ARE WE GOING TO MAKE IT TO THE MOON NOW?
I'M TOO FAR ALONG TO TURN BACK NOW, BROTHER.
HOW ARE YOU GOING TO MAKE IT BACK TO YOUR HOTEL ROOM, LOVELL?
A HOPE AND A PRAYER.
I'LL HAVE TO IMPROVISE.
RISK. ISN'T THAT WHAT WE SIGNED UP FOR?
KNOCK IT OFF, GORDO.
SERIOUSLY.
WELL, I SIGNED UP TO SEE THE SIGHTS!
WE'RE THE GUYS WHO TAKE THE RISKS!
ED, GUS AND ROGER – THEY KNEW THAT.
AND NOW THE WHOLE PROGRAM IS AT STAKE.
THEY WOULDN'T SHUT DOWN THE PROGRAM, NOT WHEN WE'RE THIS CLOSE.
AMERICA HAS AN OBLIGATION TO GO TO THE MOON NOW.
IF IT GETS SOMEONE RE-ELECTED IN WASHINGTON... THEY'LL TRY.
FOR OUR FALLEN.
THIS WON'T ALL BE IN VAIN.
THEIR SACRIFICE CAN'T BE FOR NOTHING.

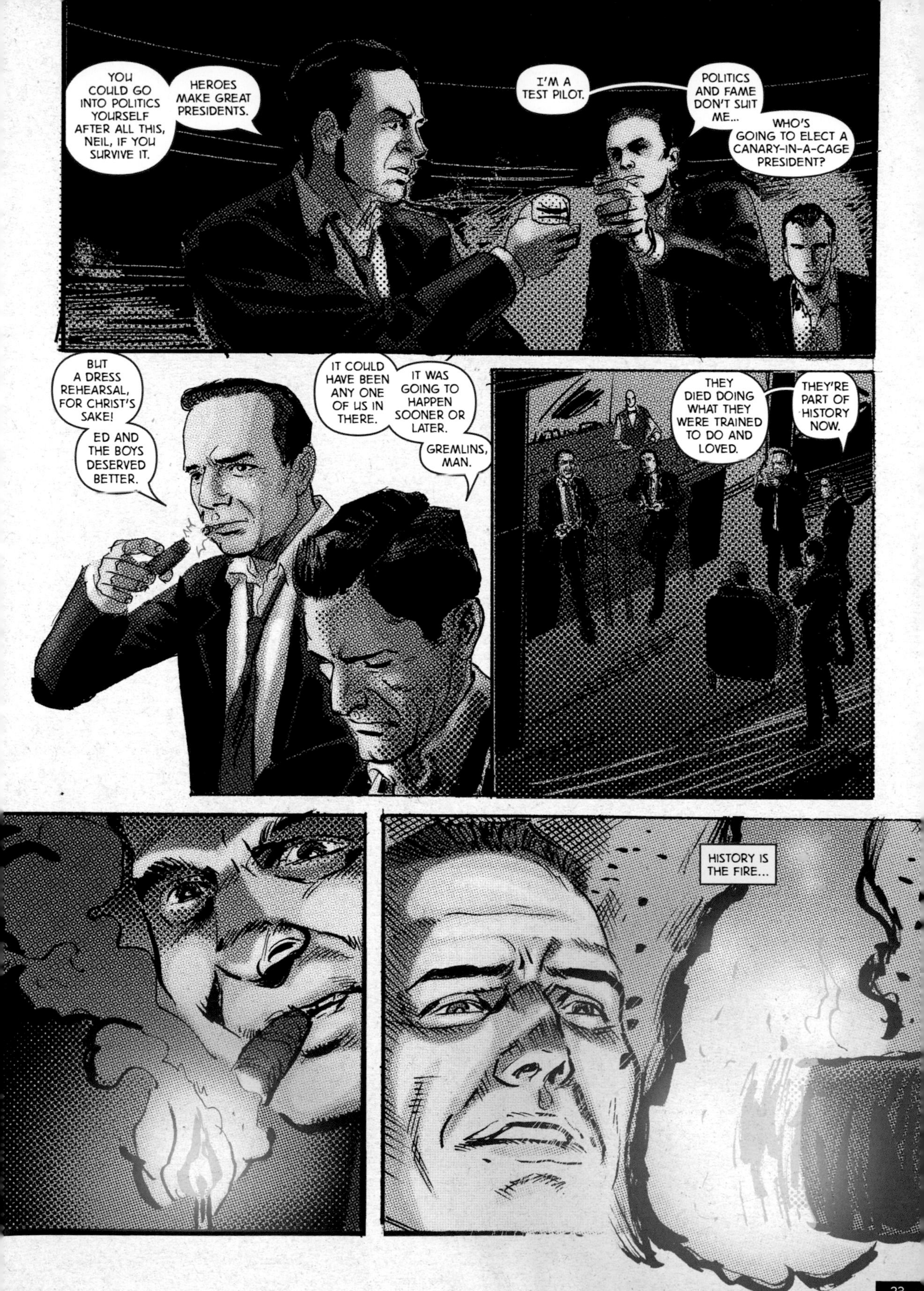
YOU COULD GO INTO POLITICS YOURSELF AFTER ALL THIS, NEIL, IF YOU SURVIVE IT.
HEROES MAKE GREAT PRESIDENTS.
I'M A TEST PILOT.
POLITICS AND FAME DON'T SUIT ME...
WHO'S GOING TO ELECT A CANARY-IN-A-CAGE PRESIDENT?
BUT A DRESS REHEARSAL, FOR CHRIST'S SAKE!
ED AND THE BOYS DESERVED BETTER.
IT COULD HAVE BEEN ANY ONE OF US IN THERE.
IT WAS GOING TO HAPPEN SOONER OR LATER.
GREMLINS, MAN.
THEY DIED DOING WHAT THEY WERE TRAINED TO DO AND LOVED.
THEY'RE PART OF HISTORY NOW.
HISTORY IS THE FIRE...

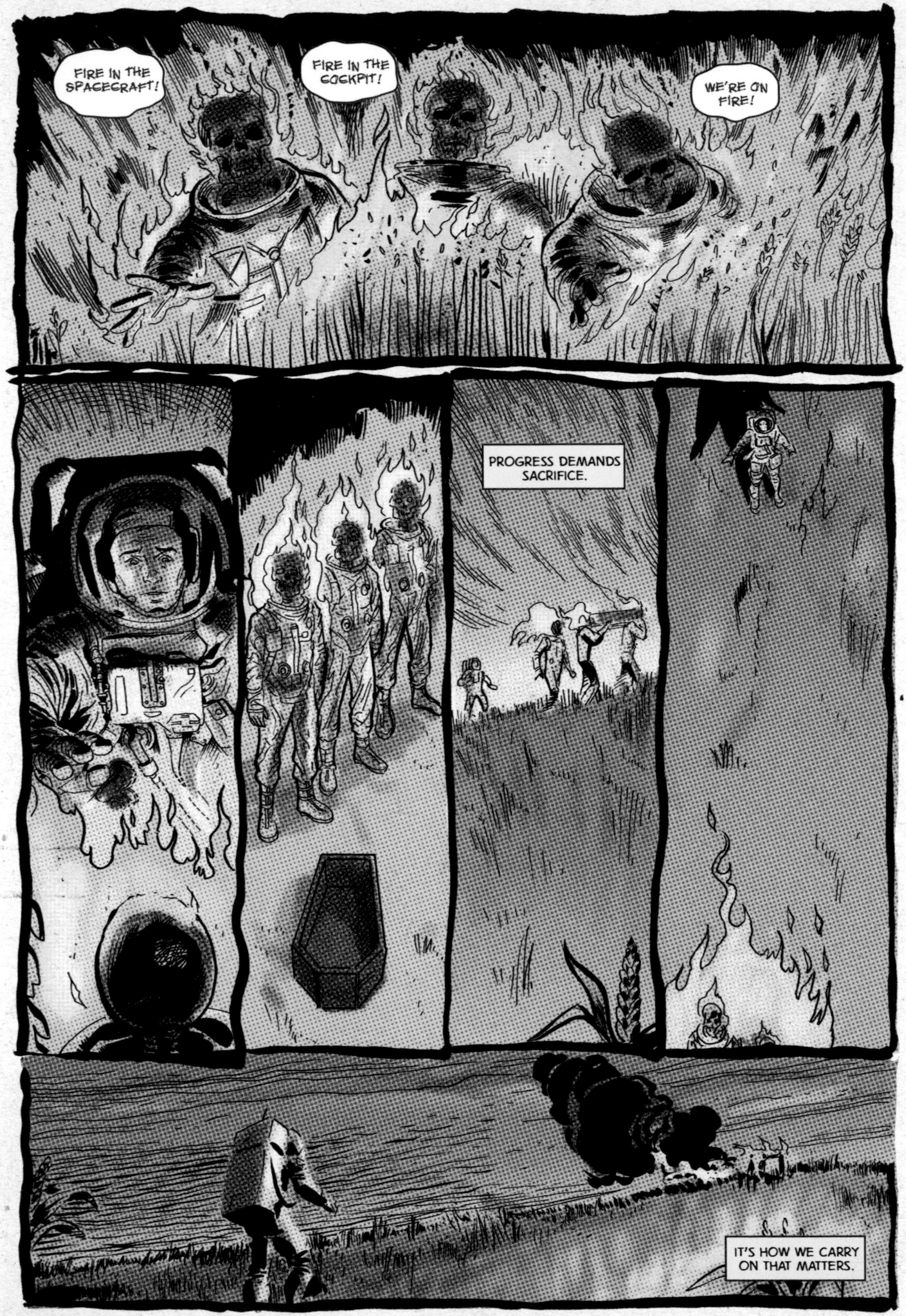
FIRE IN THE SPACECRAFT!
FIRE IN THE COCKPIT!
WE'RE ON FIRE!
PROGRESS DEMANDS SACRIFICE.
IT'S HOW WE CARRY ON THAT MATTERS.

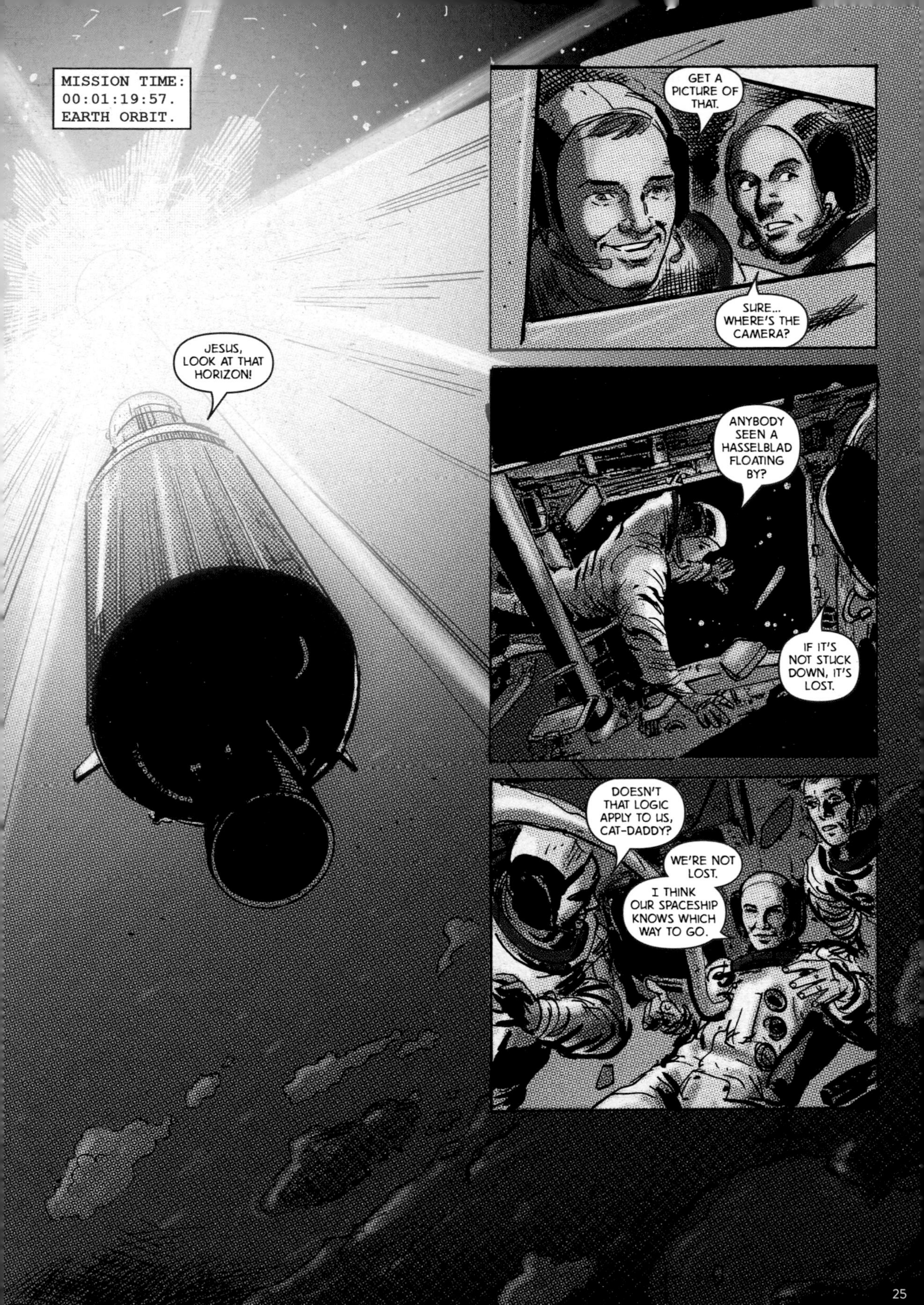
MISSION TIME:
00:01:19:57.
EARTH ORBIT.
JESUS, LOOK AT THAT HORIZON!
GET A PICTURE OF THAT.
SURE... WHERE'S THE CAMERA?
ANYBODY SEEN A HASSELBLAD FLOATING BY?
IF IT'S NOT STUCK DOWN, IT'S LOST.
DOESN'T THAT LOGIC APPLY TO US, CAT-DADDY?
WE'RE NOT LOST.
I THINK OUR SPACESHIP KNOWS WHICH WAY TO GO.

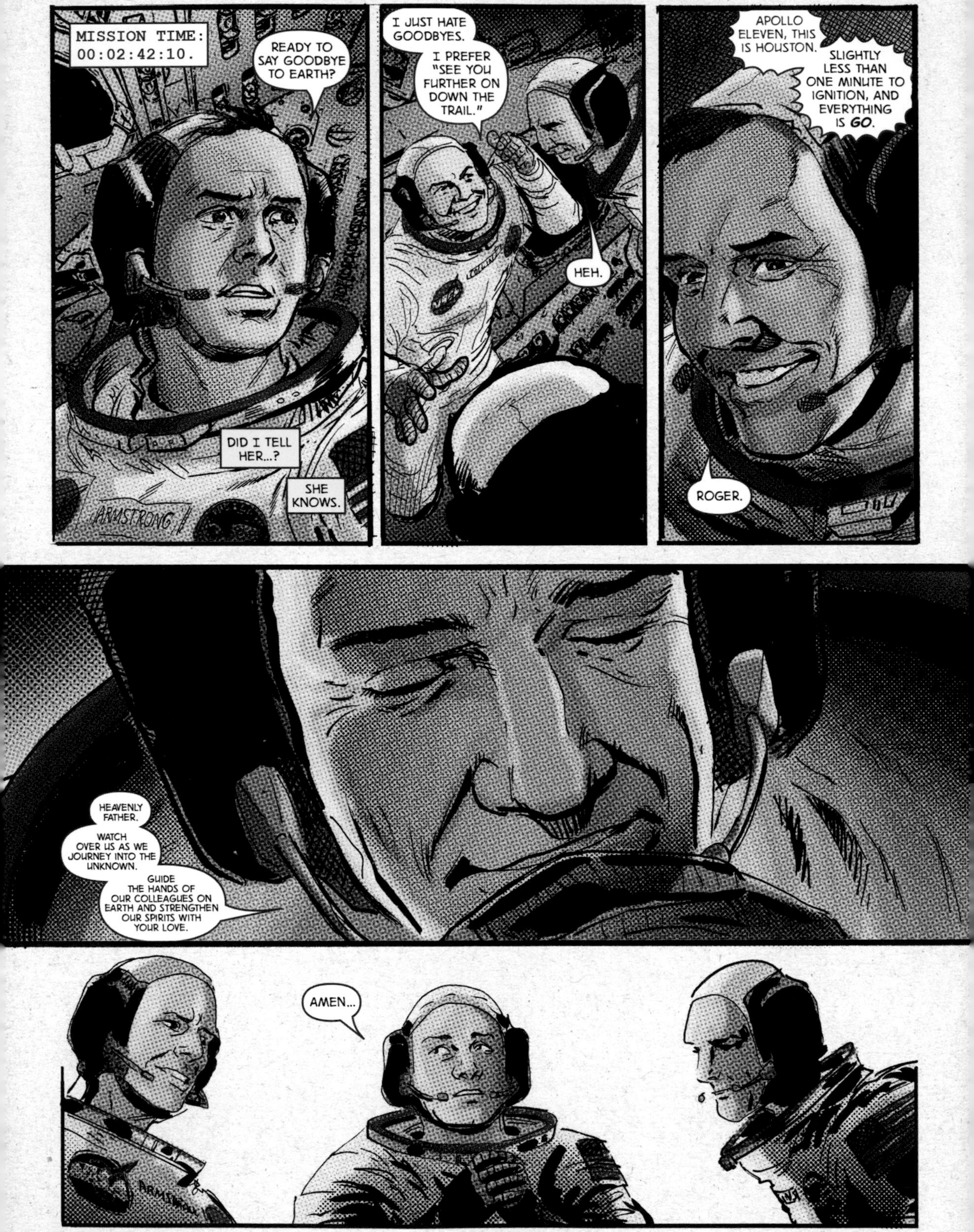
MISSION TIME: 00:02:42:10.
READY TO SAY GOODBYE TO EARTH?
DID I TELL HER...?
SHE KNOWS.
ARMSTRONG
I JUST HATE GOODBYES.
I PREFER "SEE YOU FURTHER ON DOWN THE TRAIL."
HEH.
APOLLO ELEVEN, THIS IS HOUSTON.
SLIGHTLY LESS THAN ONE MINUTE TO IGNITION, AND EVERYTHING IS **GO**.
ROGER.
HEAVENLY FATHER.
WATCH OVER US AS WE JOURNEY INTO THE UNKNOWN.
GUIDE THE HANDS OF OUR COLLEAGUES ON EARTH AND STRENGTHEN OUR SPIRITS WITH YOUR LOVE.
AMEN...

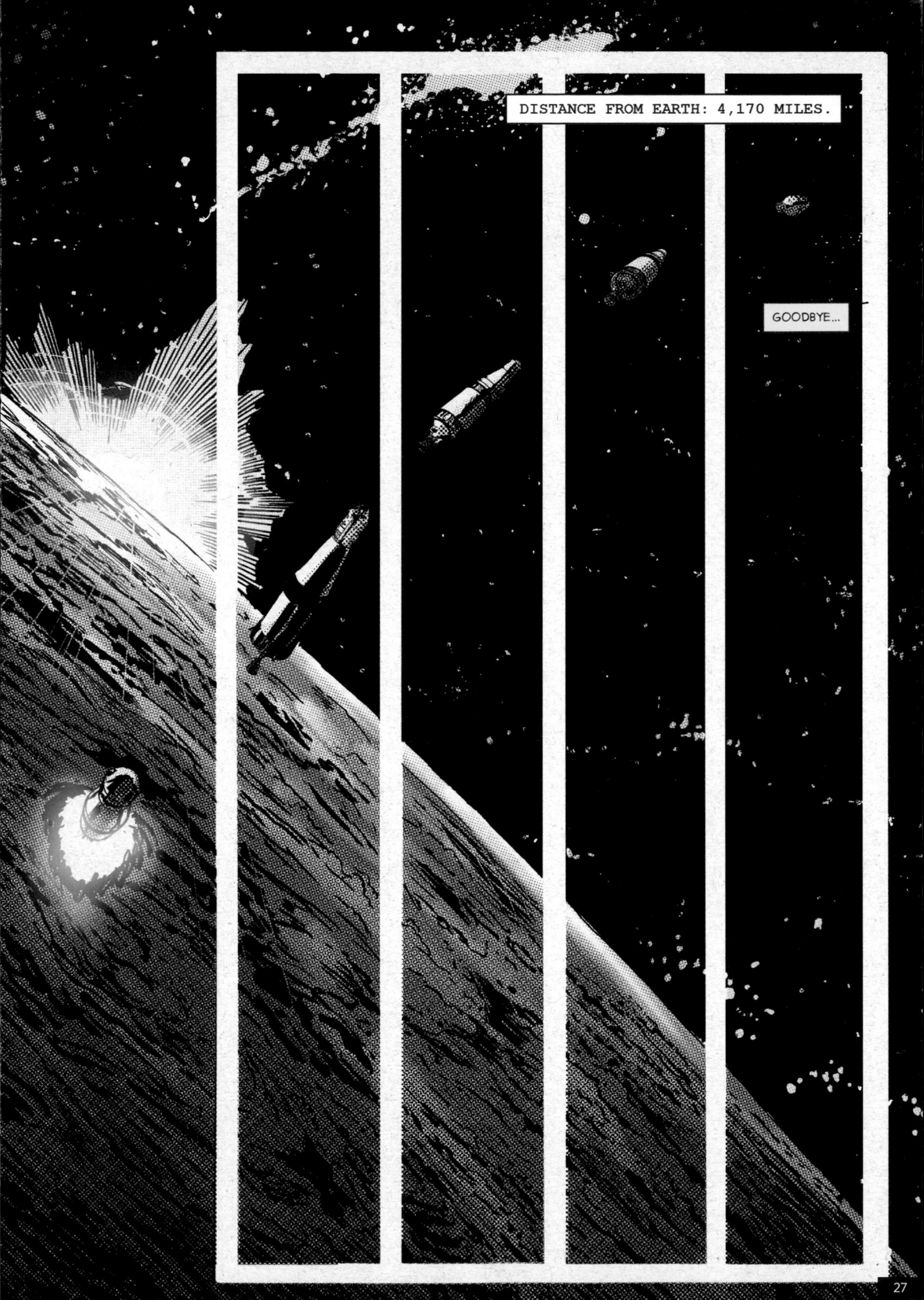
DISTANCE FROM EARTH: 4,170 MILES.
GOODBYE...

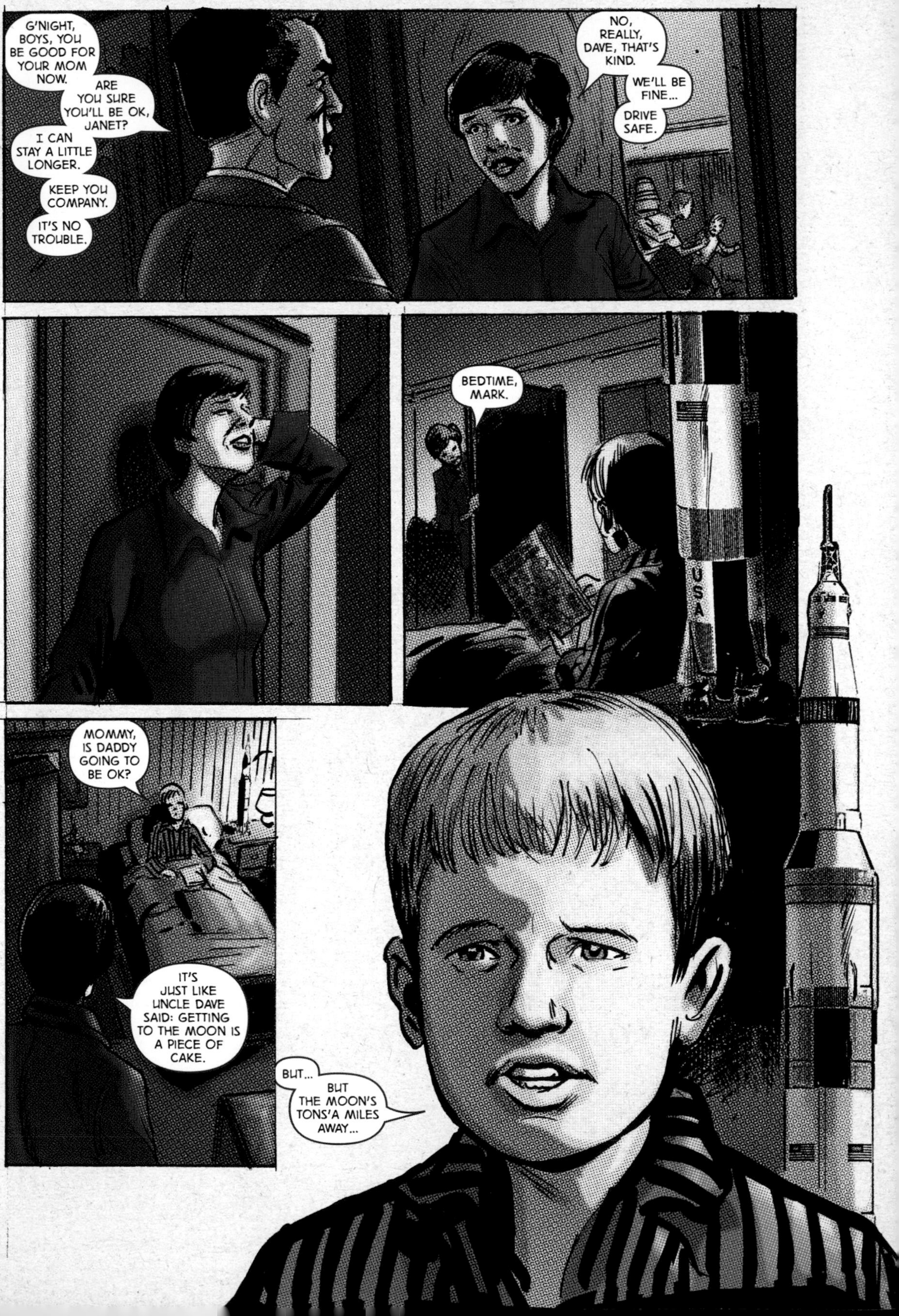
G'NIGHT, BOYS, YOU BE GOOD FOR YOUR MOM NOW.
ARE YOU SURE YOU'LL BE OK, JANET?
I CAN STAY A LITTLE LONGER.
KEEP YOU COMPANY.
IT'S NO TROUBLE.
NO, REALLY, DAVE, THAT'S KIND.
WE'LL BE FINE...
DRIVE SAFE.
BEDTIME, MARK.
USA
MOMMY, IS DADDY GOING TO BE OK?
IT'S JUST LIKE UNCLE DAVE SAID: GETTING TO THE MOON IS A PIECE OF CAKE.
BUT...
BUT THE MOON'S TONS'A MILES AWAY...

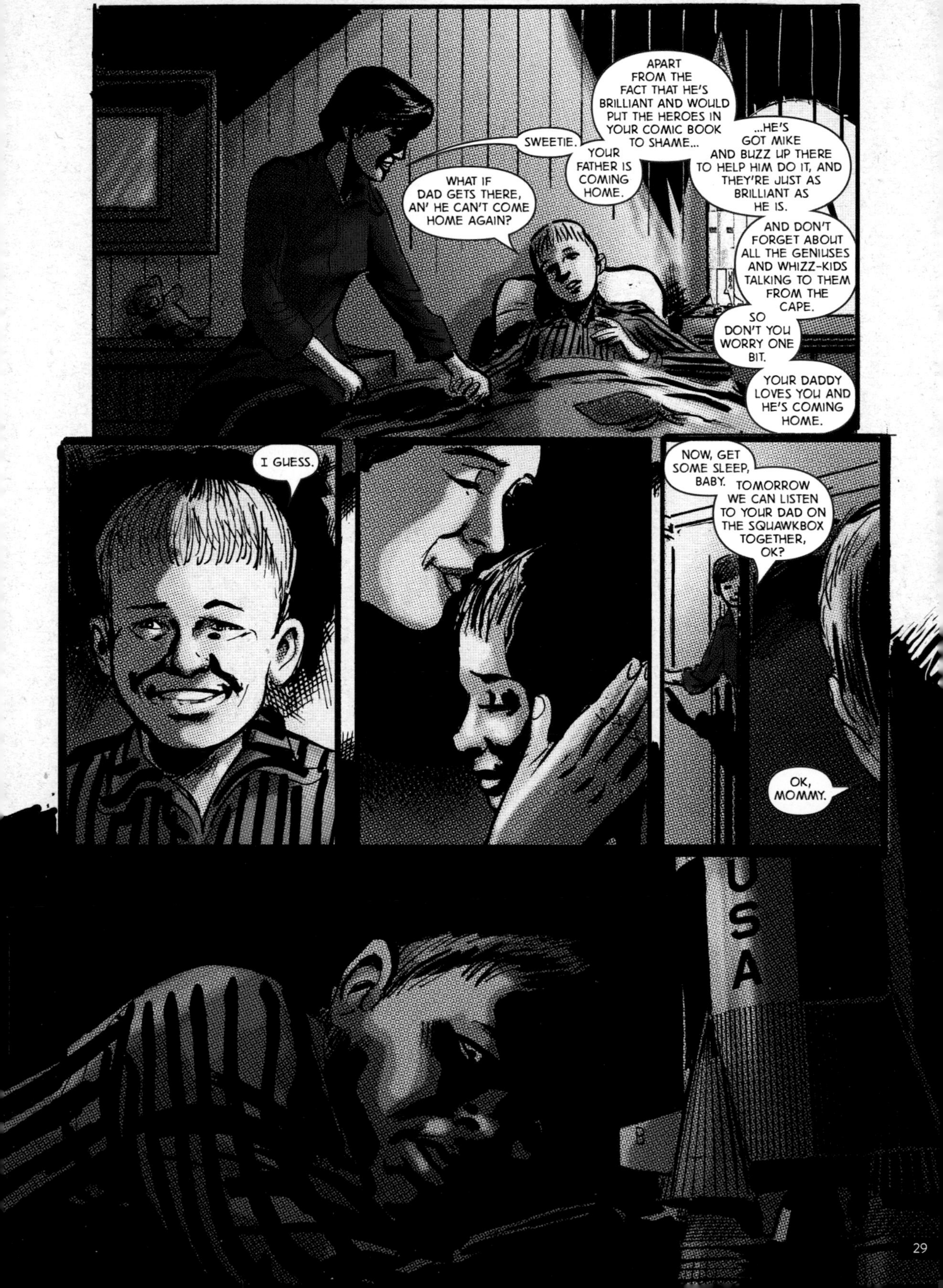
WHAT IF DAD GETS THERE, AN' HE CAN'T COME HOME AGAIN?
SWEETIE.
YOUR FATHER IS COMING HOME.
APART FROM THE FACT THAT HE'S BRILLIANT AND WOULD PUT THE HEROES IN YOUR COMIC BOOK TO SHAME...
...HE'S GOT MIKE AND BUZZ UP THERE TO HELP HIM DO IT, AND THEY'RE JUST AS BRILLIANT AS HE IS.
AND DON'T FORGET ABOUT ALL THE GENIUSES AND WHIZZ-KIDS TALKING TO THEM FROM THE CAPE.
SO DON'T YOU WORRY ONE BIT.
YOUR DADDY LOVES YOU AND HE'S COMING HOME.
I GUESS.
NOW, GET SOME SLEEP, BABY.
TOMORROW WE CAN LISTEN TO YOUR DAD ON THE SQUAWKBOX TOGETHER, OK?
OK, MOMMY.
USA

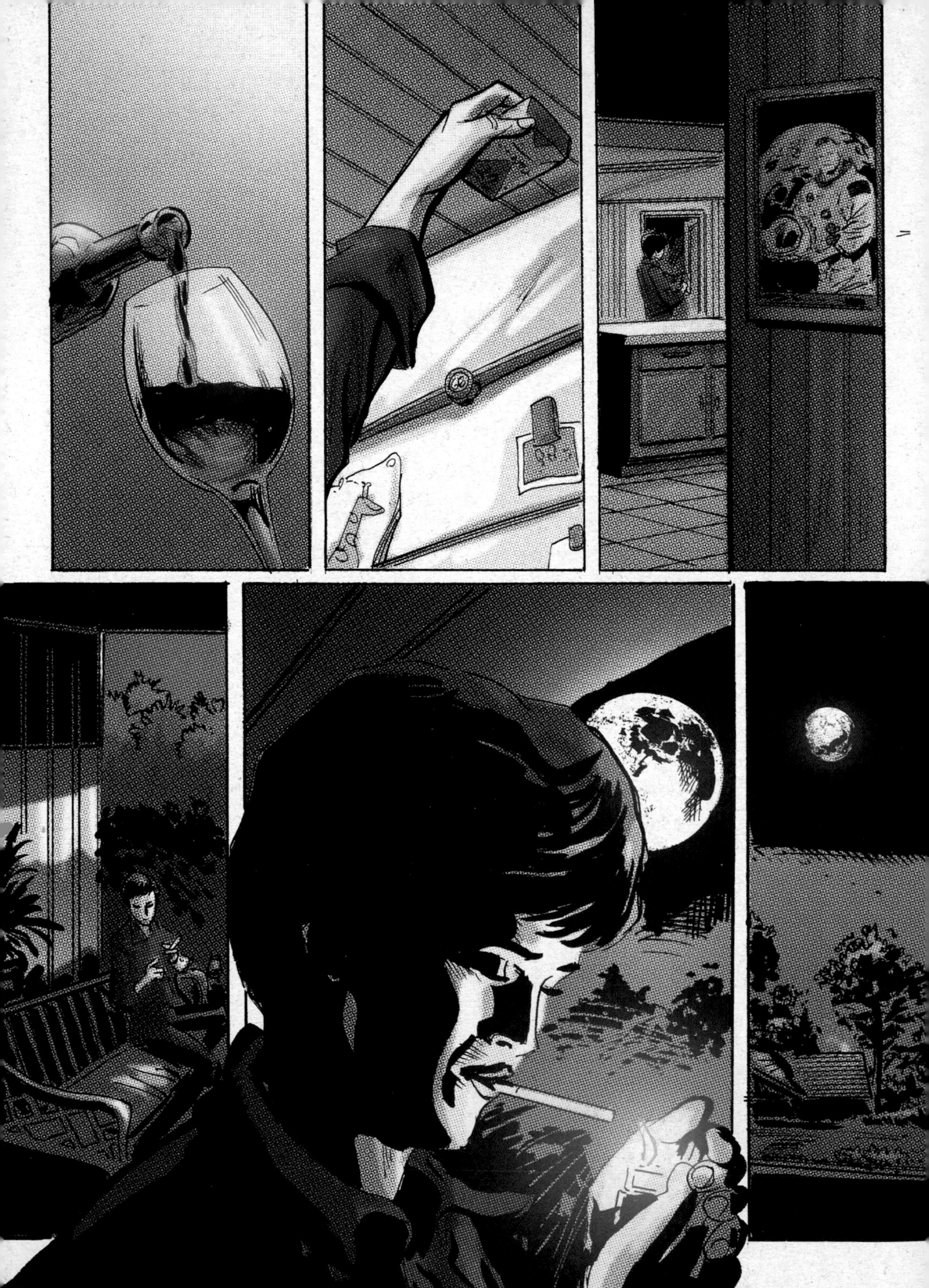

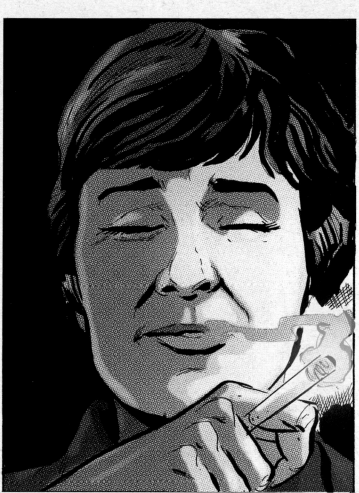

HE'S COMING
BACK.

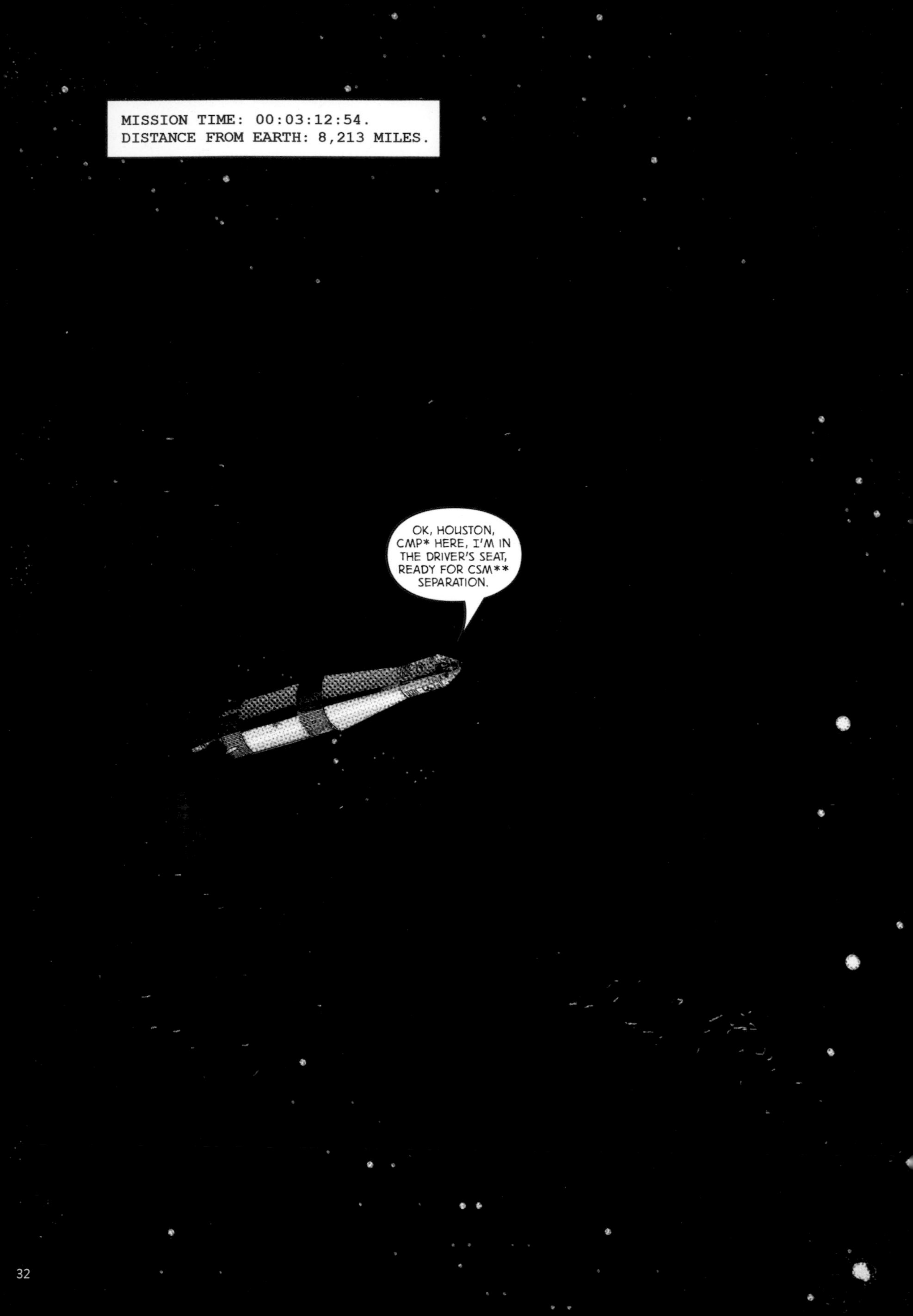
MISSION TIME: 00:03:12:54.
DISTANCE FROM EARTH: 8,213 MILES.
OK, HOUSTON, CMP* HERE, I'M IN THE DRIVER'S SEAT, READY FOR CSM** SEPARATION.

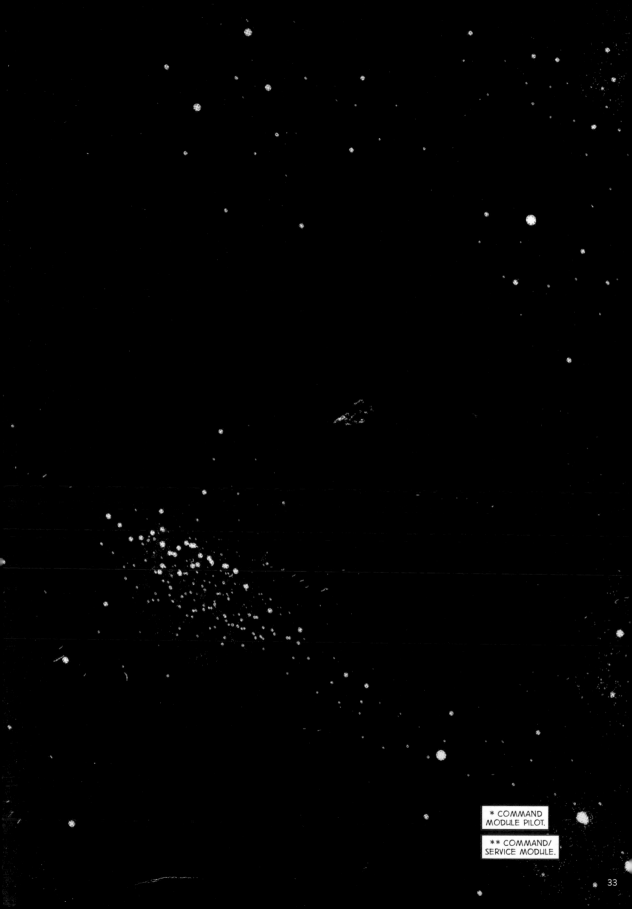

* COMMAND
MODULE PILOT.

** COMMAND/
SERVICE MODULE.

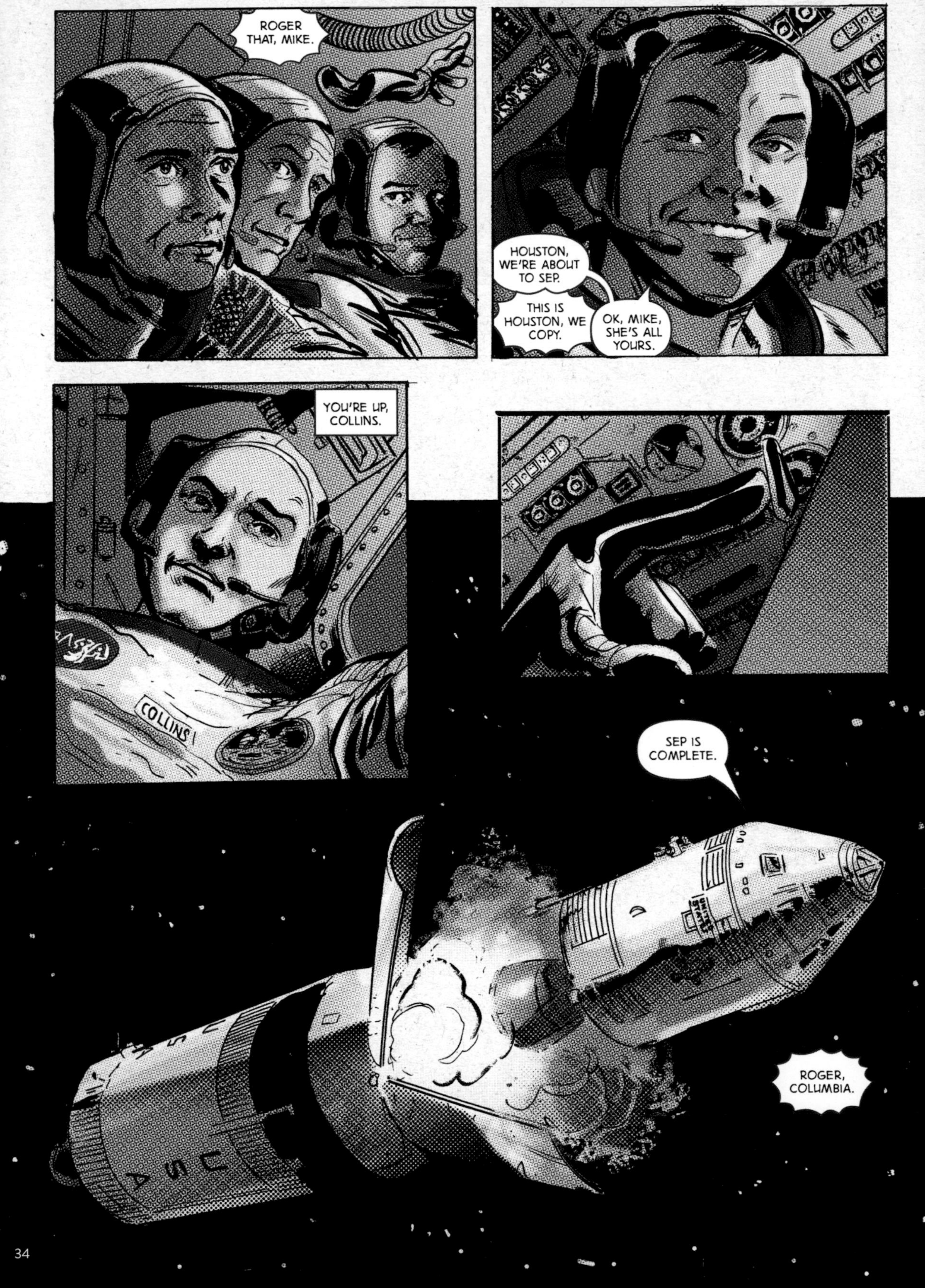
ROGER THAT, MIKE.
HOUSTON, WE'RE ABOUT TO SEP.
THIS IS HOUSTON, WE COPY.
OK, MIKE, SHE'S ALL YOURS.
YOU'RE UP, COLLINS.
COLLINS
SEP IS COMPLETE.
ROGER, COLUMBIA.

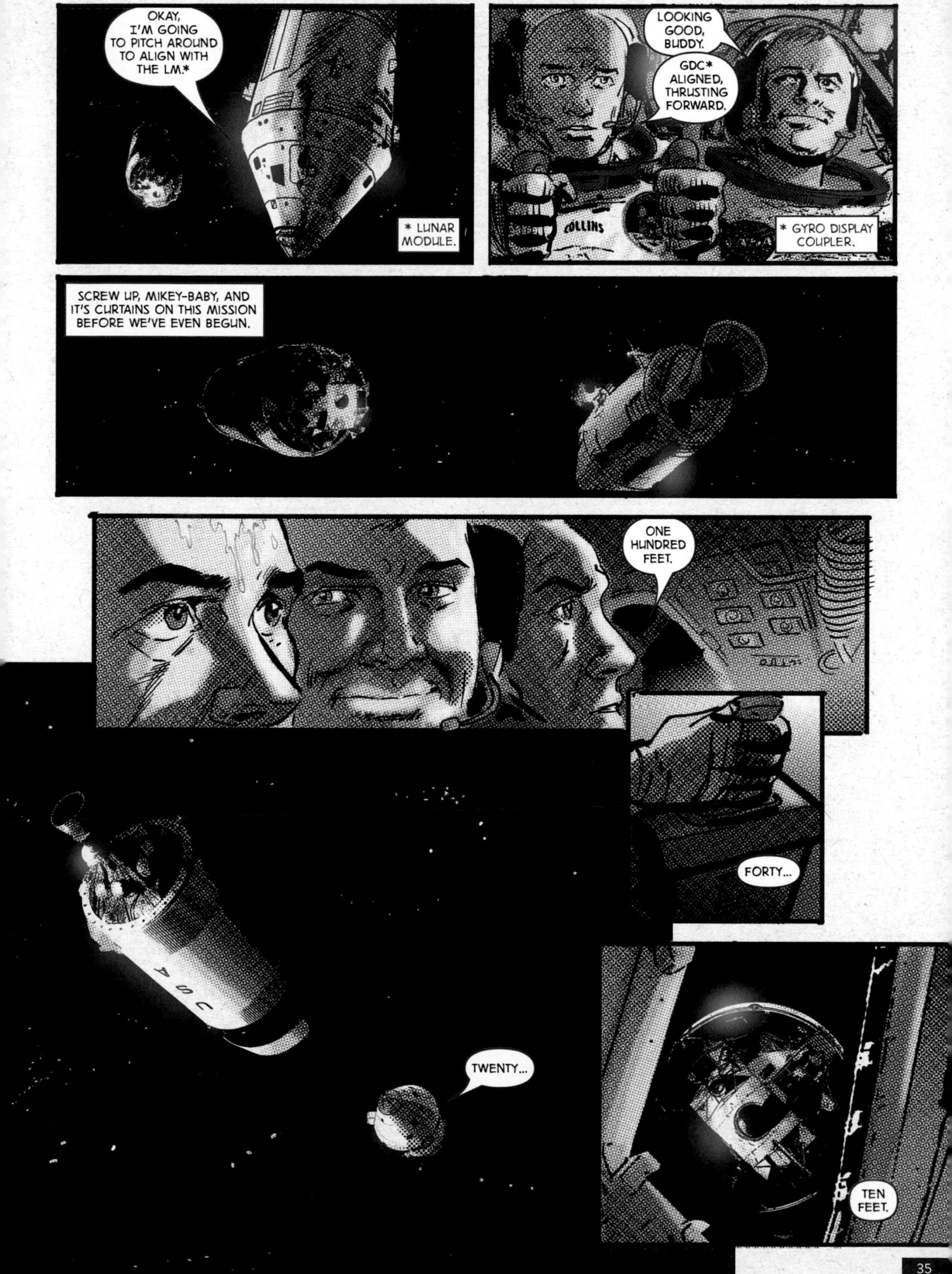
OKAY, I'M GOING TO PITCH AROUND TO ALIGN WITH THE LM.*
* LUNAR MODULE.
LOOKING GOOD, BUDDY.
GDC* ALIGNED, THRUSTING FORWARD.
COLLINS
* GYRO DISPLAY COUPLER.
SCREW UP, MIKEY-BABY, AND IT'S CURTAINS ON THIS MISSION BEFORE WE'VE EVEN BEGUN.
ONE HUNDRED FEET.
FORTY...
TWENTY...
TEN FEET.

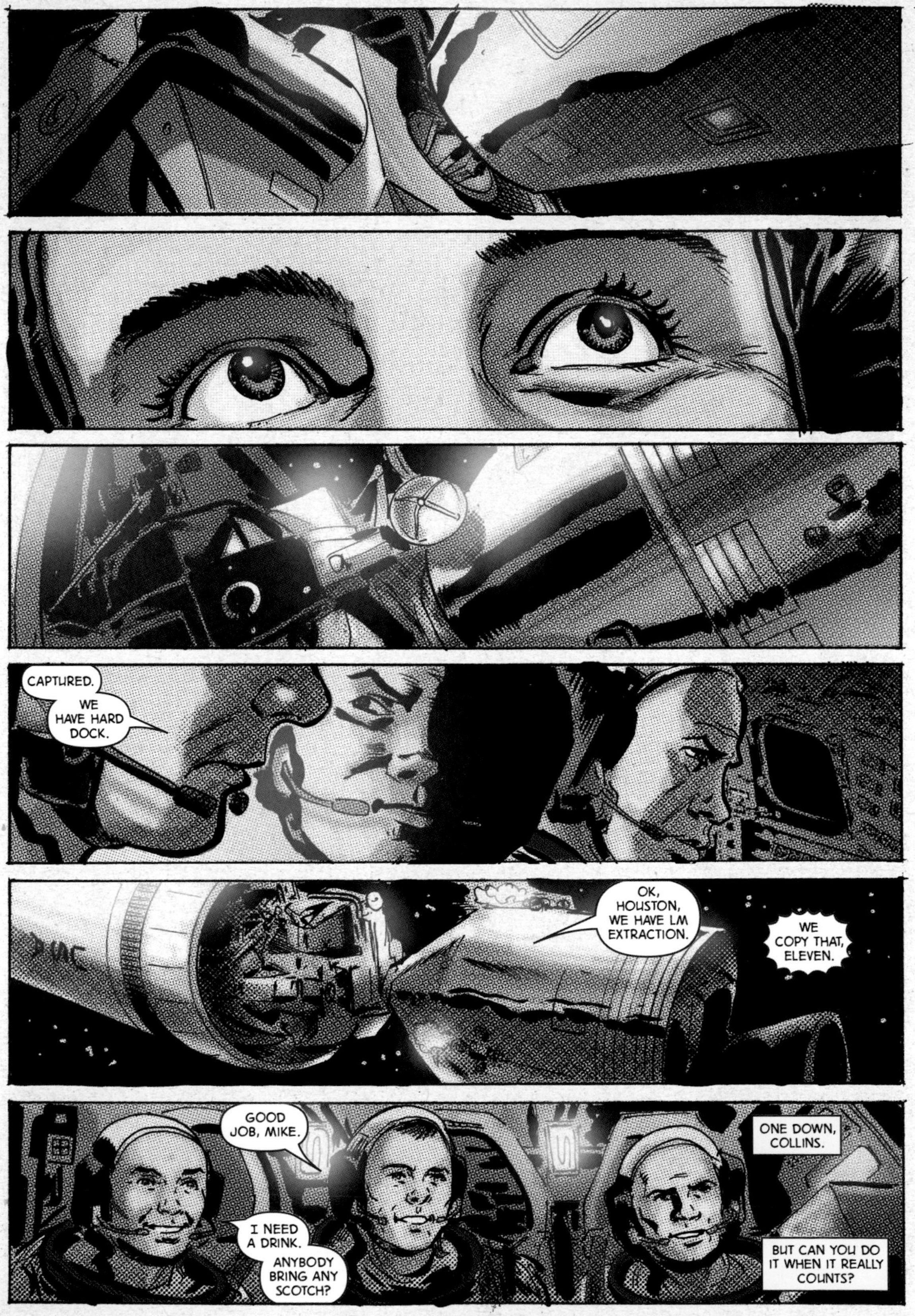
CAPTURED.
WE HAVE HARD DOCK.
OK, HOUSTON, WE HAVE LM EXTRACTION.
WE COPY THAT, ELEVEN.
GOOD JOB, MIKE.
I NEED A DRINK.
ANYBODY BRING ANY SCOTCH?
ONE DOWN, COLLINS.
BUT CAN YOU DO IT WHEN IT REALLY COUNTS?

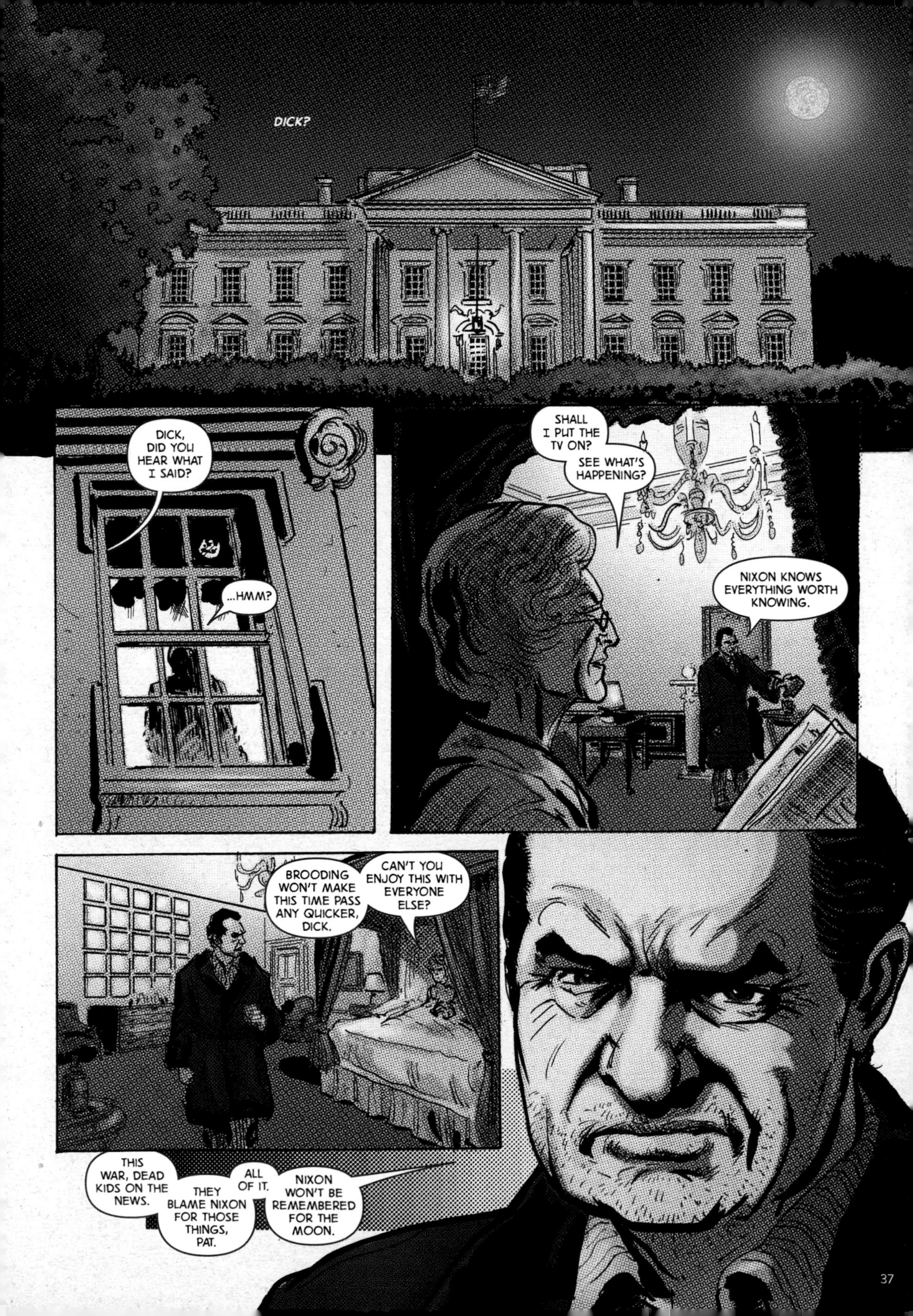
DICK?
DICK, DID YOU HEAR WHAT I SAID?
...HMM?
SHALL I PUT THE TV ON?
SEE WHAT'S HAPPENING?
NIXON KNOWS EVERYTHING WORTH KNOWING.
BROODING WON'T MAKE THIS TIME PASS ANY QUICKER, DICK.
CAN'T YOU ENJOY THIS WITH EVERYONE ELSE?
THIS WAR, DEAD KIDS ON THE NEWS.
THEY BLAME NIXON FOR THOSE THINGS, PAT.
ALL OF IT.
NIXON WON'T BE REMEMBERED FOR THE MOON.

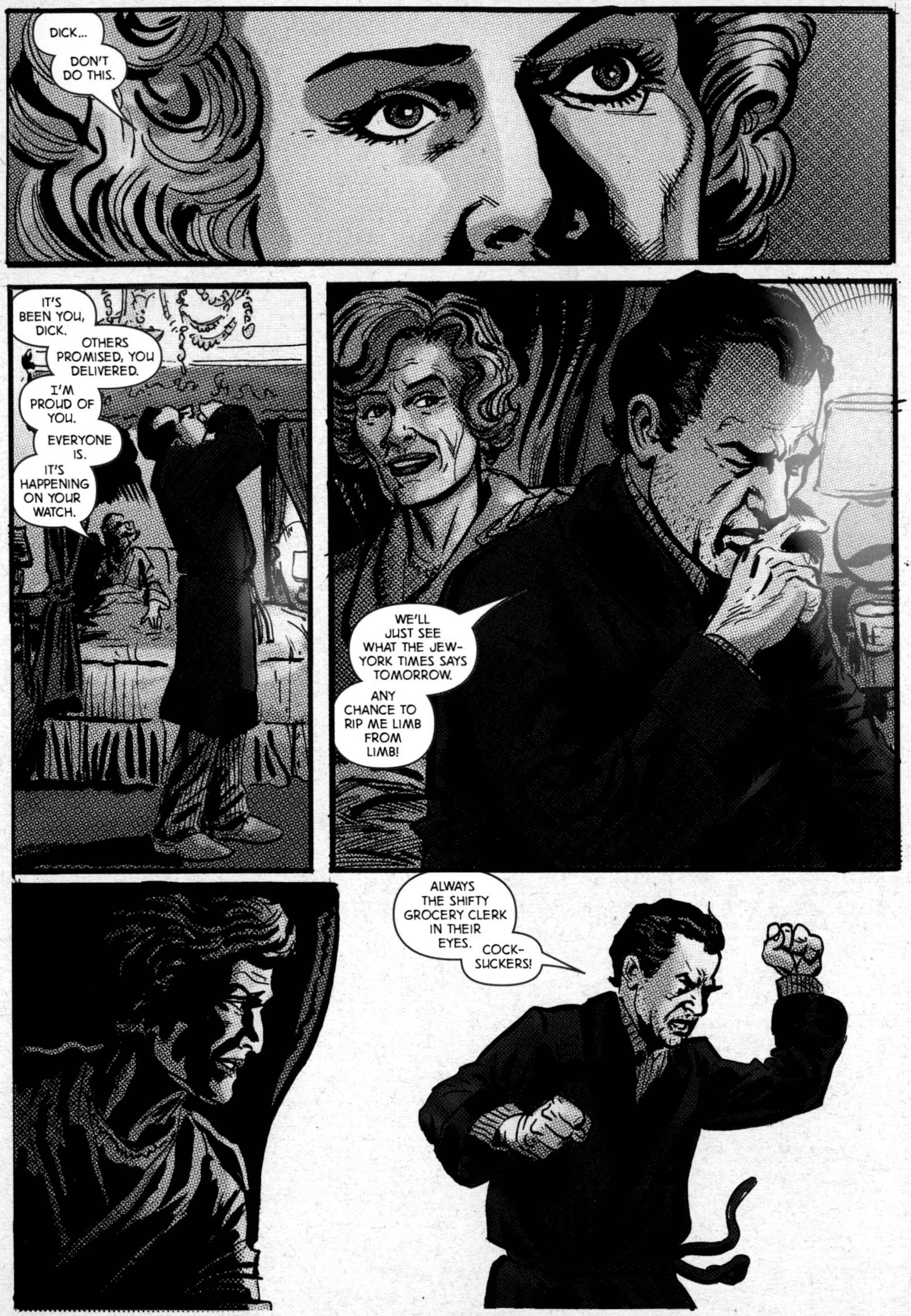
DICK... DON'T DO THIS.
IT'S BEEN YOU, DICK.
OTHERS PROMISED, YOU DELIVERED.
I'M PROUD OF YOU.
EVERYONE IS.
IT'S HAPPENING ON YOUR WATCH.
WE'LL JUST SEE WHAT THE JEW-YORK TIMES SAYS TOMORROW.
ANY CHANCE TO RIP ME LIMB FROM LIMB!
ALWAYS THE SHIFTY GROCERY CLERK IN THEIR EYES.
COCK-SUCKERS!

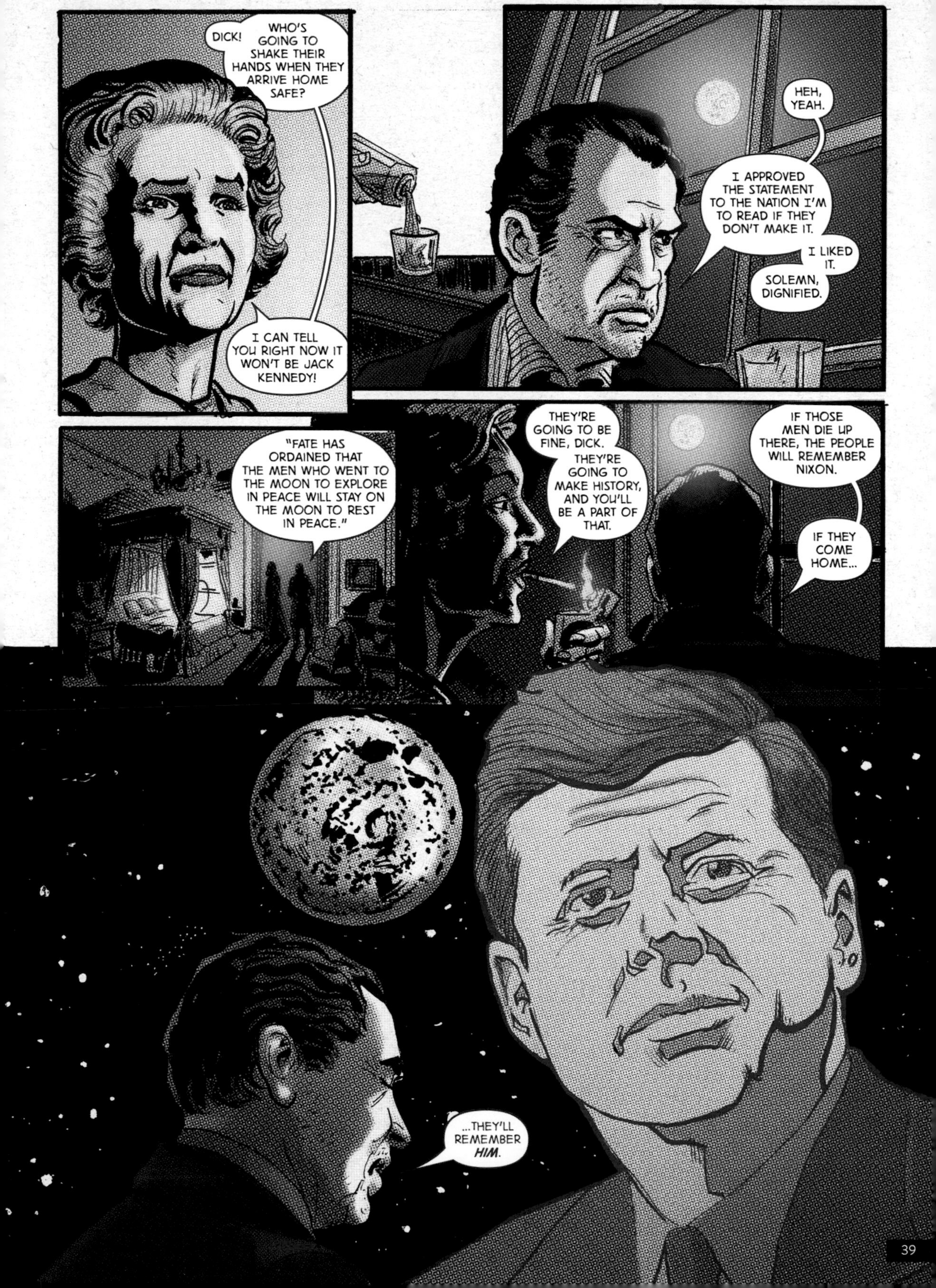
DICK!
WHO'S GOING TO SHAKE THEIR HANDS WHEN THEY ARRIVE HOME SAFE?
I CAN TELL YOU RIGHT NOW IT WON'T BE JACK KENNEDY!
HEH, YEAH.
I APPROVED THE STATEMENT TO THE NATION I'M TO READ IF THEY DON'T MAKE IT.
I LIKED IT.
SOLEMN, DIGNIFIED.
"FATE HAS ORDAINED THAT THE MEN WHO WENT TO THE MOON TO EXPLORE IN PEACE WILL STAY ON THE MOON TO REST IN PEACE."
THEY'RE GOING TO BE FINE, DICK.
THEY'RE GOING TO MAKE HISTORY, AND YOU'LL BE A PART OF THAT.
IF THOSE MEN DIE UP THERE, THE PEOPLE WILL REMEMBER NIXON.
IF THEY COME HOME...
...THEY'LL REMEMBER *HIM*.

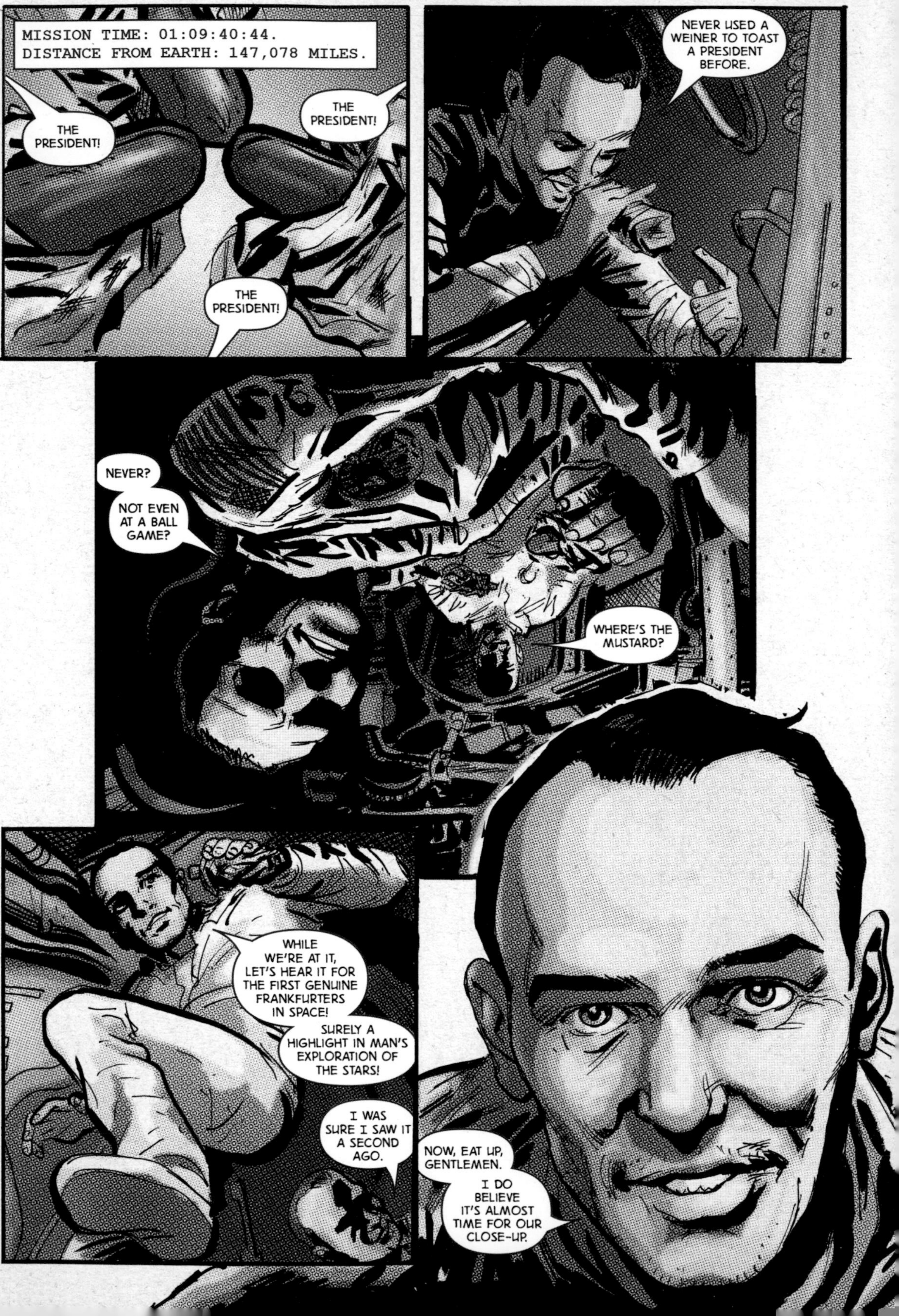
MISSION TIME: 01:09:40:44.
DISTANCE FROM EARTH: 147,078 MILES.
THE PRESIDENT!
THE PRESIDENT!
THE PRESIDENT!
NEVER USED A WEINER TO TOAST A PRESIDENT BEFORE.
NEVER?
NOT EVEN AT A BALL GAME?
WHERE'S THE MUSTARD?
WHILE WE'RE AT IT, LET'S HEAR IT FOR THE FIRST GENUINE FRANKFURTERS IN SPACE!
SURELY A HIGHLIGHT IN MAN'S EXPLORATION OF THE STARS!
I WAS SURE I SAW IT A SECOND AGO.
NOW, EAT UP, GENTLEMEN.
I DO BELIEVE IT'S ALMOST TIME FOR OUR CLOSE-UP.

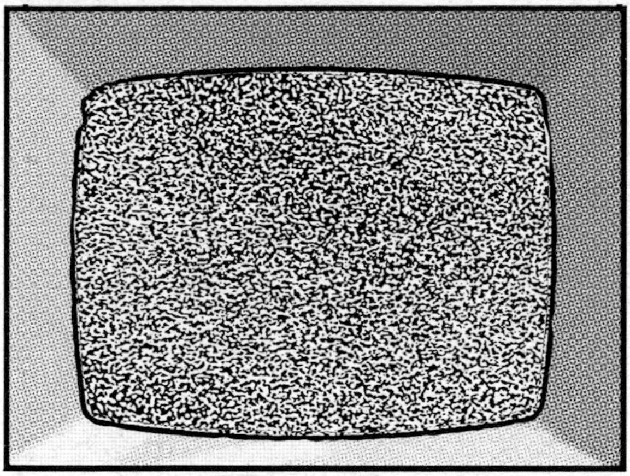

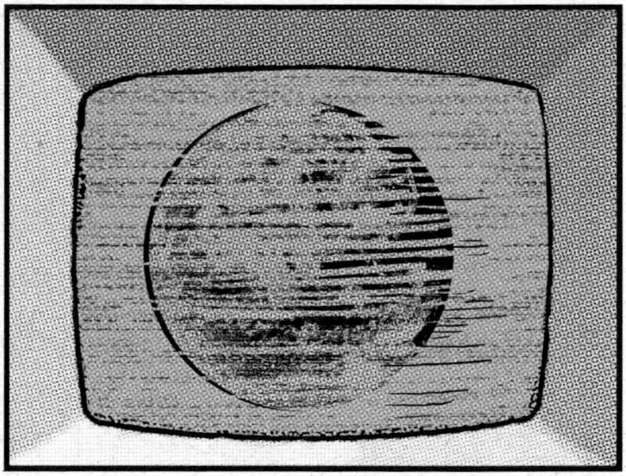

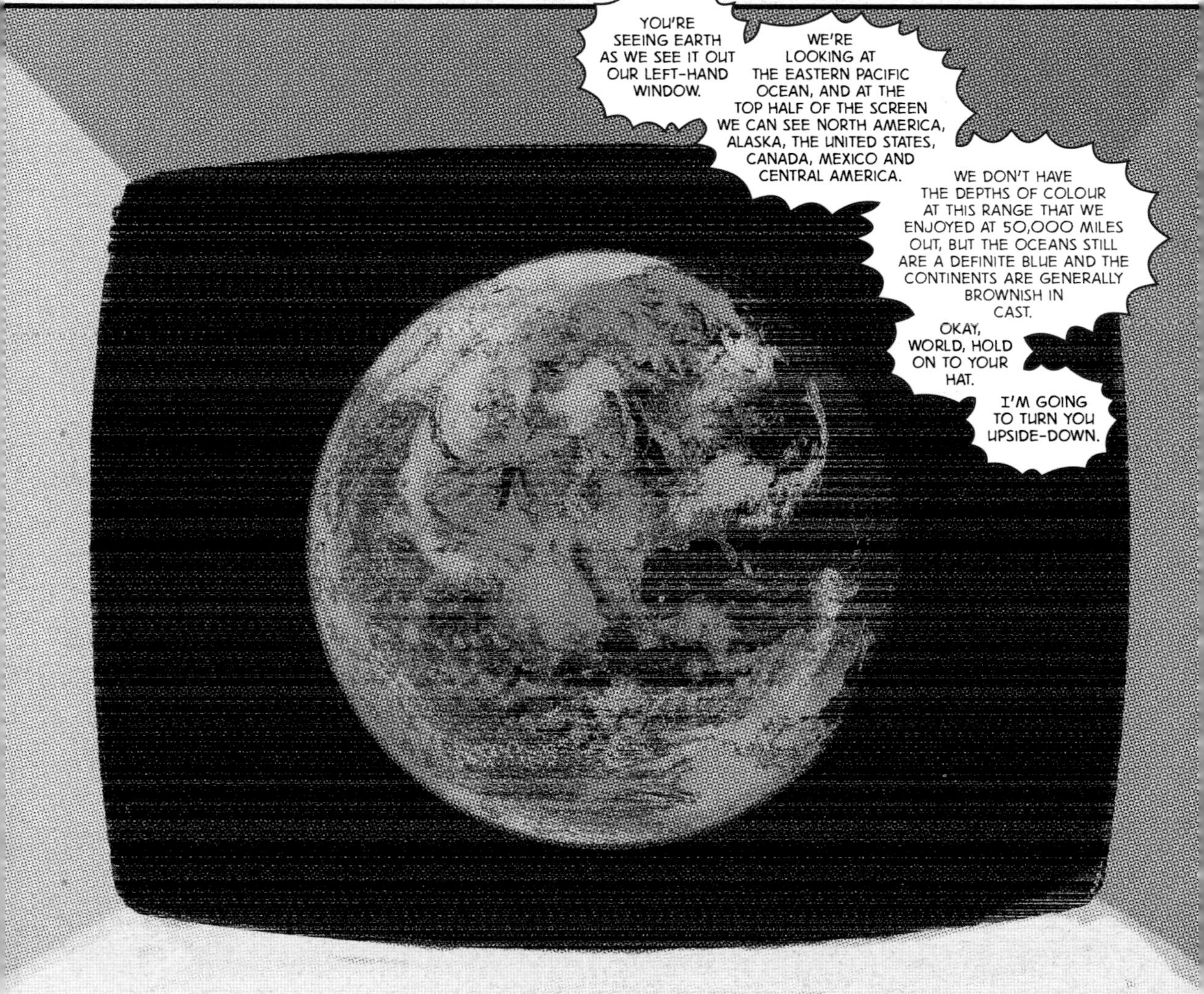
YOU'RE SEEING EARTH AS WE SEE IT OUT OUR LEFT-HAND WINDOW.
WE'RE LOOKING AT THE EASTERN PACIFIC OCEAN, AND AT THE TOP HALF OF THE SCREEN WE CAN SEE NORTH AMERICA, ALASKA, THE UNITED STATES, CANADA, MEXICO AND CENTRAL AMERICA.
WE DON'T HAVE THE DEPTHS OF COLOUR AT THIS RANGE THAT WE ENJOYED AT 50,000 MILES OUT, BUT THE OCEANS STILL ARE A DEFINITE BLUE AND THE CONTINENTS ARE GENERALLY BROWNISH IN CAST.
OKAY, WORLD, HOLD ON TO YOUR HAT.
I'M GOING TO TURN YOU UPSIDE-DOWN.

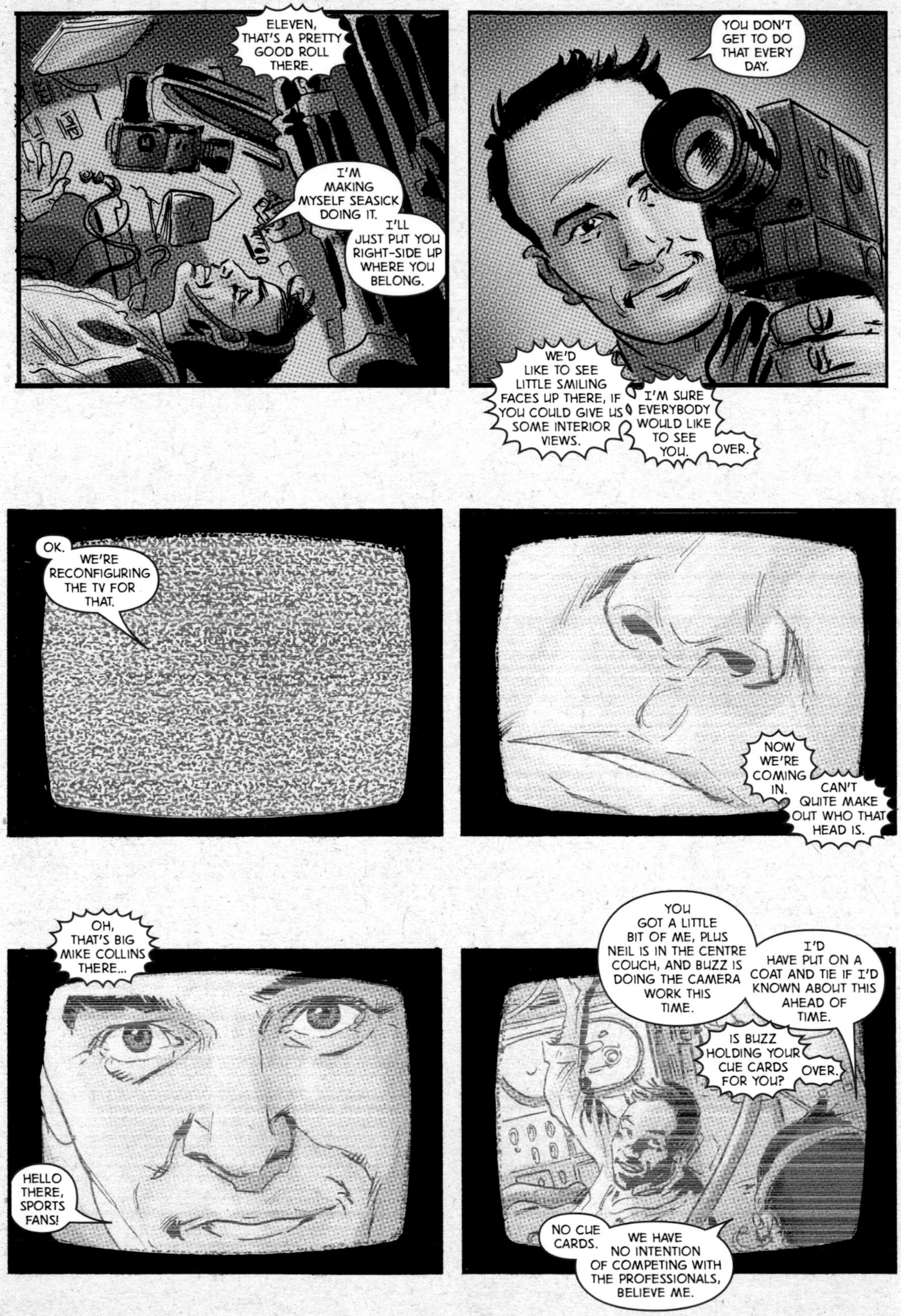
ELEVEN, THAT'S A PRETTY GOOD ROLL THERE.
I'M MAKING MYSELF SEASICK DOING IT.
I'LL JUST PUT YOU RIGHT-SIDE UP WHERE YOU BELONG.
YOU DON'T GET TO DO THAT EVERY DAY.
WE'D LIKE TO SEE LITTLE SMILING FACES UP THERE, IF YOU COULD GIVE US SOME INTERIOR VIEWS.
I'M SURE EVERYBODY WOULD LIKE TO SEE YOU.
OVER.
OK.
WE'RE RECONFIGURING THE TV FOR THAT.
NOW WE'RE COMING IN.
CAN'T QUITE MAKE OUT WHO THAT HEAD IS.
OH, THAT'S BIG MIKE COLLINS THERE...
HELLO THERE, SPORTS FANS!
YOU GOT A LITTLE BIT OF ME, PLUS NEIL IS IN THE CENTRE COUCH, AND BUZZ IS DOING THE CAMERA WORK THIS TIME.
I'D HAVE PUT ON A COAT AND TIE IF I'D KNOWN ABOUT THIS AHEAD OF TIME.
IS BUZZ HOLDING YOUR CUE CARDS FOR YOU?
OVER.
NO CUE CARDS.
WE HAVE NO INTENTION OF COMPETING WITH THE PROFESSIONALS, BELIEVE ME.

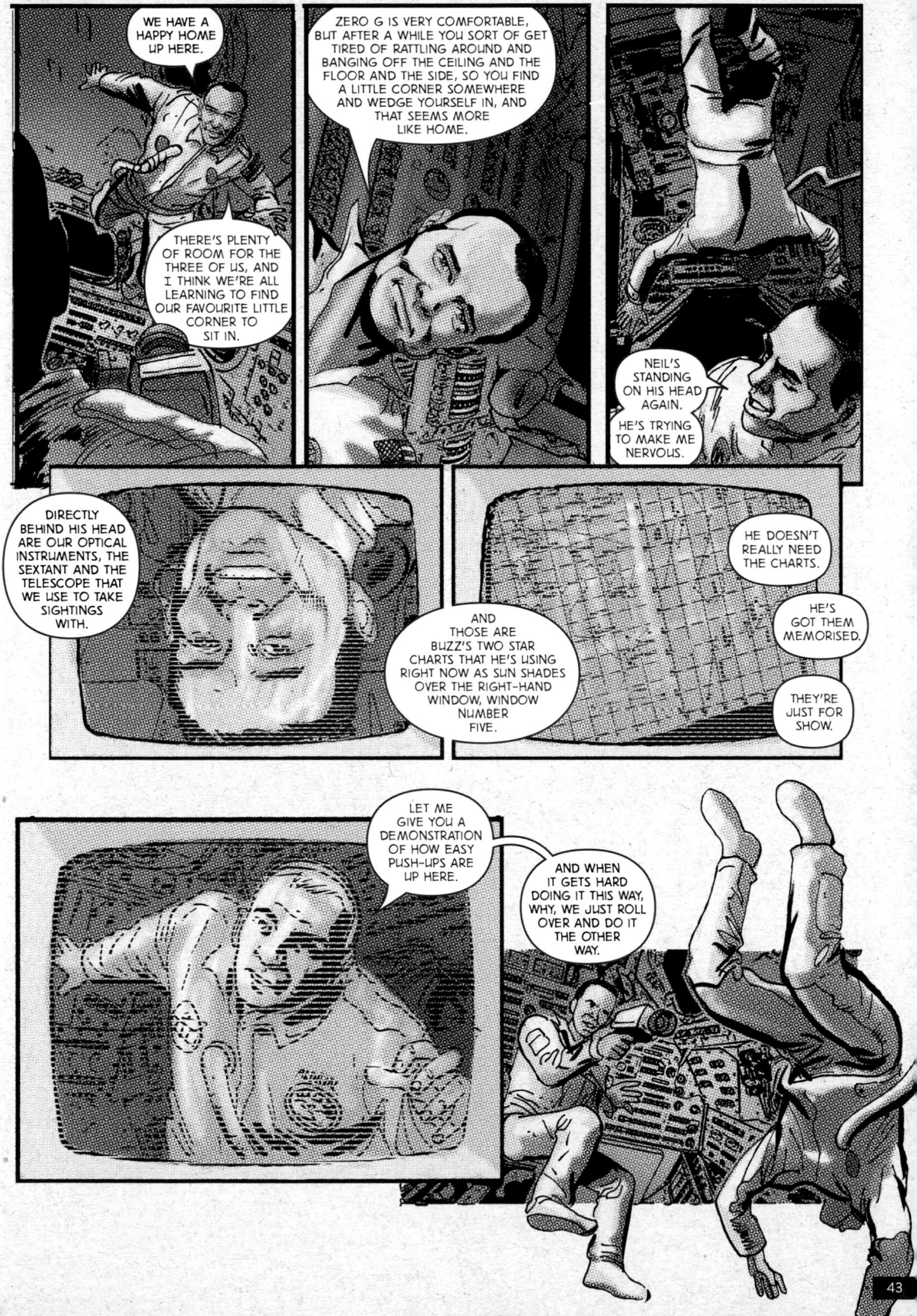
WE HAVE A HAPPY HOME UP HERE.
THERE'S PLENTY OF ROOM FOR THE THREE OF US, AND I THINK WE'RE ALL LEARNING TO FIND OUR FAVOURITE LITTLE CORNER TO SIT IN.
ZERO G IS VERY COMFORTABLE, BUT AFTER A WHILE YOU SORT OF GET TIRED OF RATTLING AROUND AND BANGING OFF THE CEILING AND THE FLOOR AND THE SIDE, SO YOU FIND A LITTLE CORNER SOMEWHERE AND WEDGE YOURSELF IN, AND THAT SEEMS MORE LIKE HOME.
NEIL'S STANDING ON HIS HEAD AGAIN.
HE'S TRYING TO MAKE ME NERVOUS.
DIRECTLY BEHIND HIS HEAD ARE OUR OPTICAL INSTRUMENTS, THE SEXTANT AND THE TELESCOPE THAT WE USE TO TAKE SIGHTINGS WITH.
AND THOSE ARE BUZZ'S TWO STAR CHARTS THAT HE'S USING RIGHT NOW AS SUN SHADES OVER THE RIGHT-HAND WINDOW, WINDOW NUMBER FIVE.
HE DOESN'T REALLY NEED THE CHARTS.
HE'S GOT THEM MEMORISED.
THEY'RE JUST FOR SHOW.
LET ME GIVE YOU A DEMONSTRATION OF HOW EASY PUSH-UPS ARE UP HERE.
AND WHEN IT GETS HARD DOING IT THIS WAY, WHY, WE JUST ROLL OVER AND DO IT THE OTHER WAY.

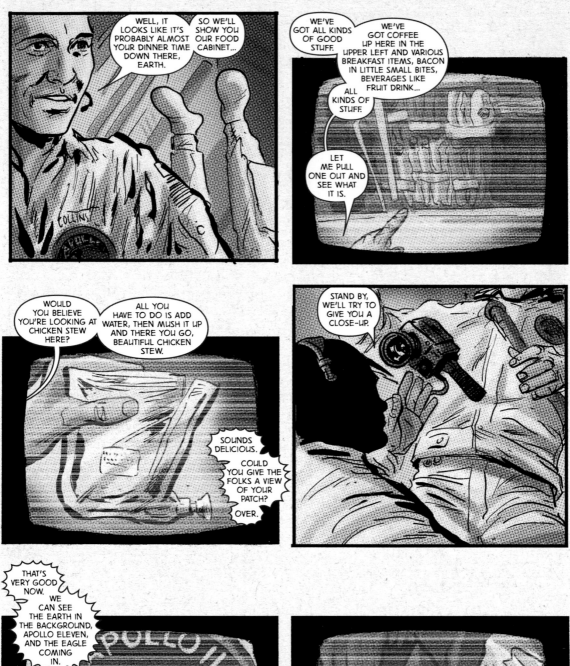

WELL, IT LOOKS LIKE IT'S PROBABLY ALMOST YOUR DINNER TIME DOWN THERE, EARTH.

SO WE'LL SHOW YOU OUR FOOD CABINET...

WE'VE GOT ALL KINDS OF GOOD STUFF.

WE'VE GOT COFFEE UP HERE IN THE UPPER LEFT AND VARIOUS BREAKFAST ITEMS, BACON IN LITTLE SMALL BITES, BEVERAGES LIKE FRUIT DRINK...

ALL KINDS OF STUFF.

LET ME PULL ONE OUT AND SEE WHAT IT IS.

WOULD YOU BELIEVE YOU'RE LOOKING AT CHICKEN STEW HERE?

ALL YOU HAVE TO DO IS ADD WATER, THEN MUSH IT UP AND THERE YOU GO, BEAUTIFUL CHICKEN STEW.

SOUNDS DELICIOUS.

COULD YOU GIVE THE FOLKS A VIEW OF YOUR PATCH?

OVER.

STAND BY, WE'LL TRY TO GIVE YOU A CLOSE-UP.

THAT'S VERY GOOD NOW. WE CAN SEE THE EARTH IN THE BACKGROUND, APOLLO ELEVEN, AND THE EAGLE COMING IN.

IT'S PROBABLY PRETTY HARD TO SEE THE OLIVE BRANCH, ISN'T IT?

ROGER.

IT IS.

WELL, THAT'S WHAT HE HAS IN HIS TALONS, AN OLIVE BRANCH.

IT LOOKS GREAT, ELEVEN.

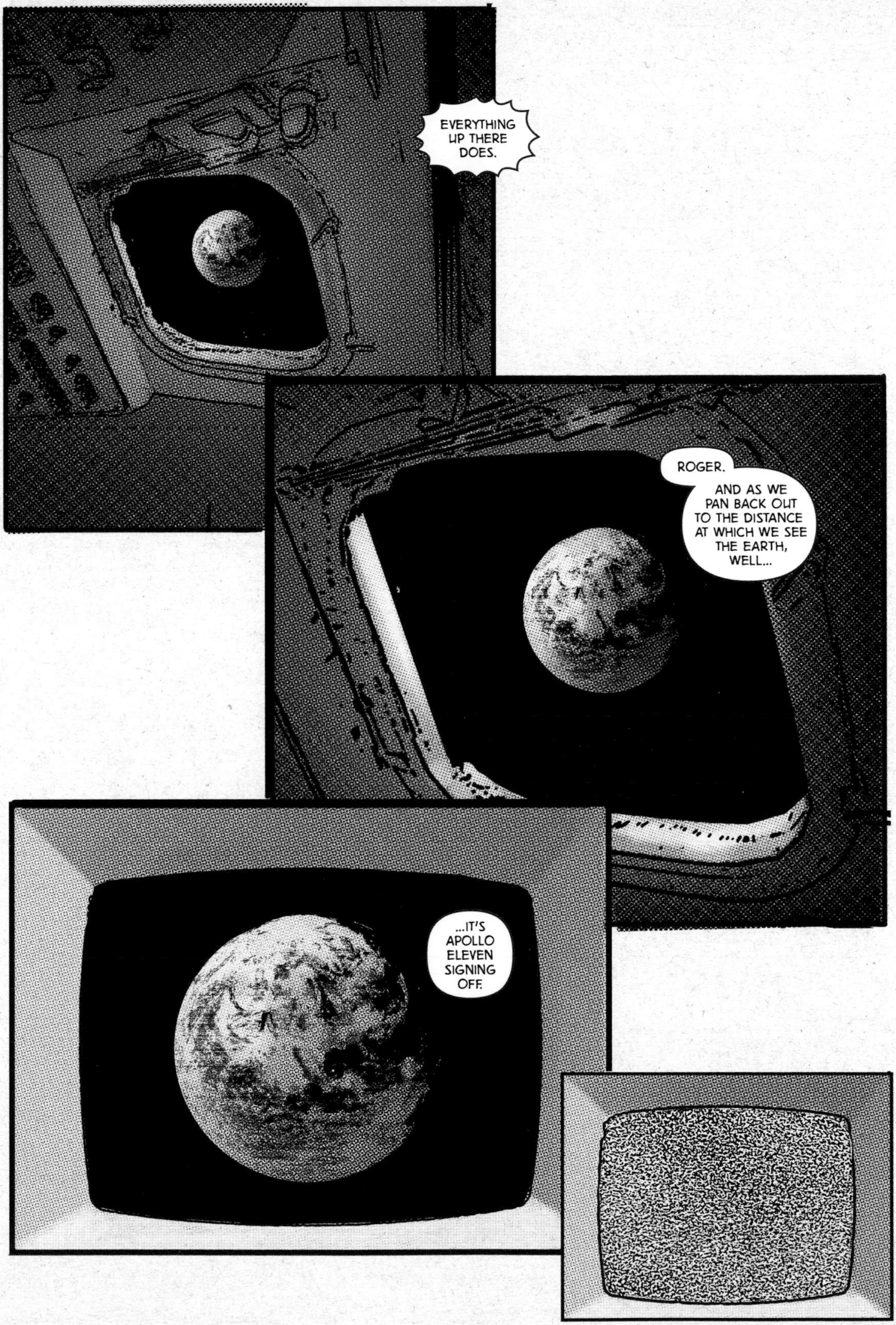
EVERYTHING UP THERE DOES.
ROGER.
AND AS WE PAN BACK OUT TO THE DISTANCE AT WHICH WE SEE THE EARTH, WELL...
...IT'S APOLLO ELEVEN SIGNING OFF.

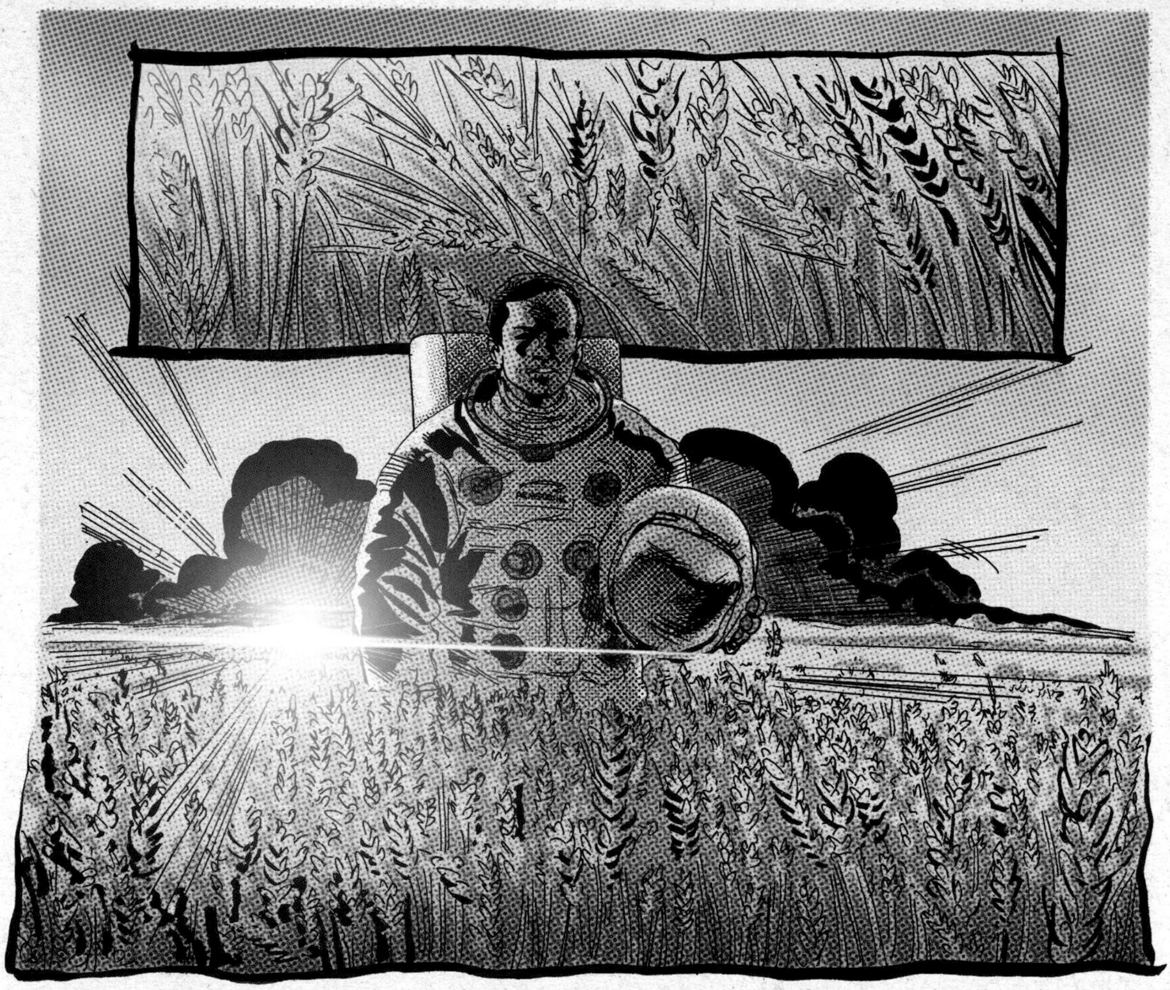

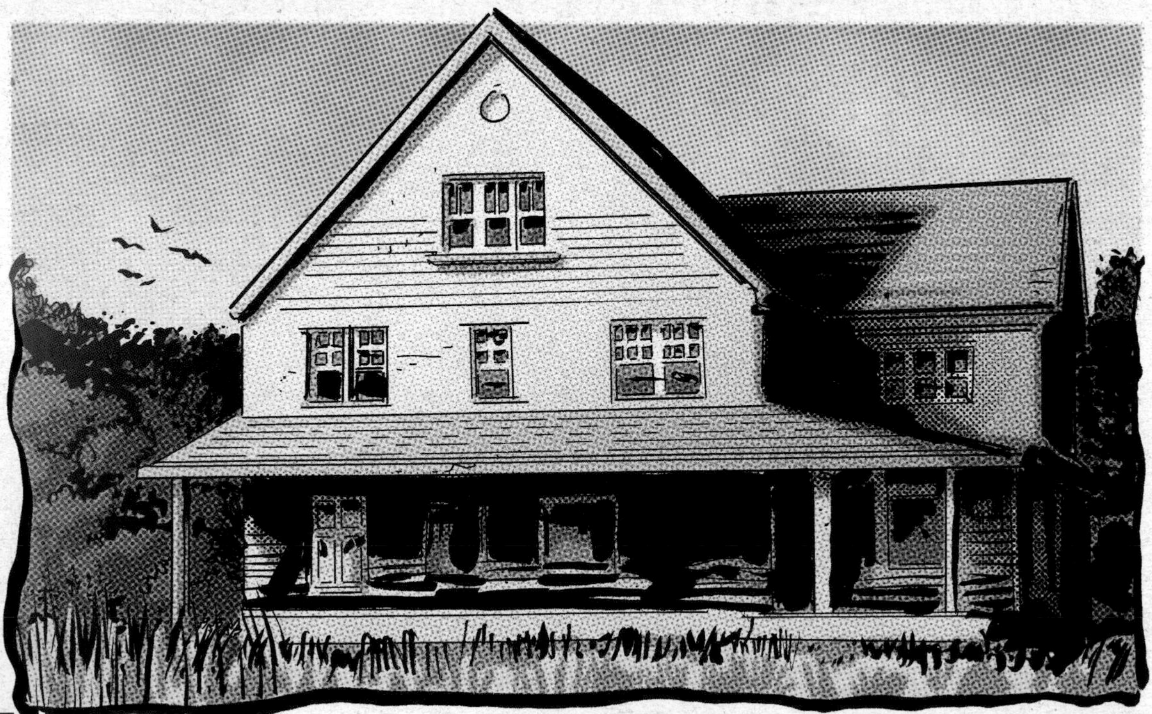

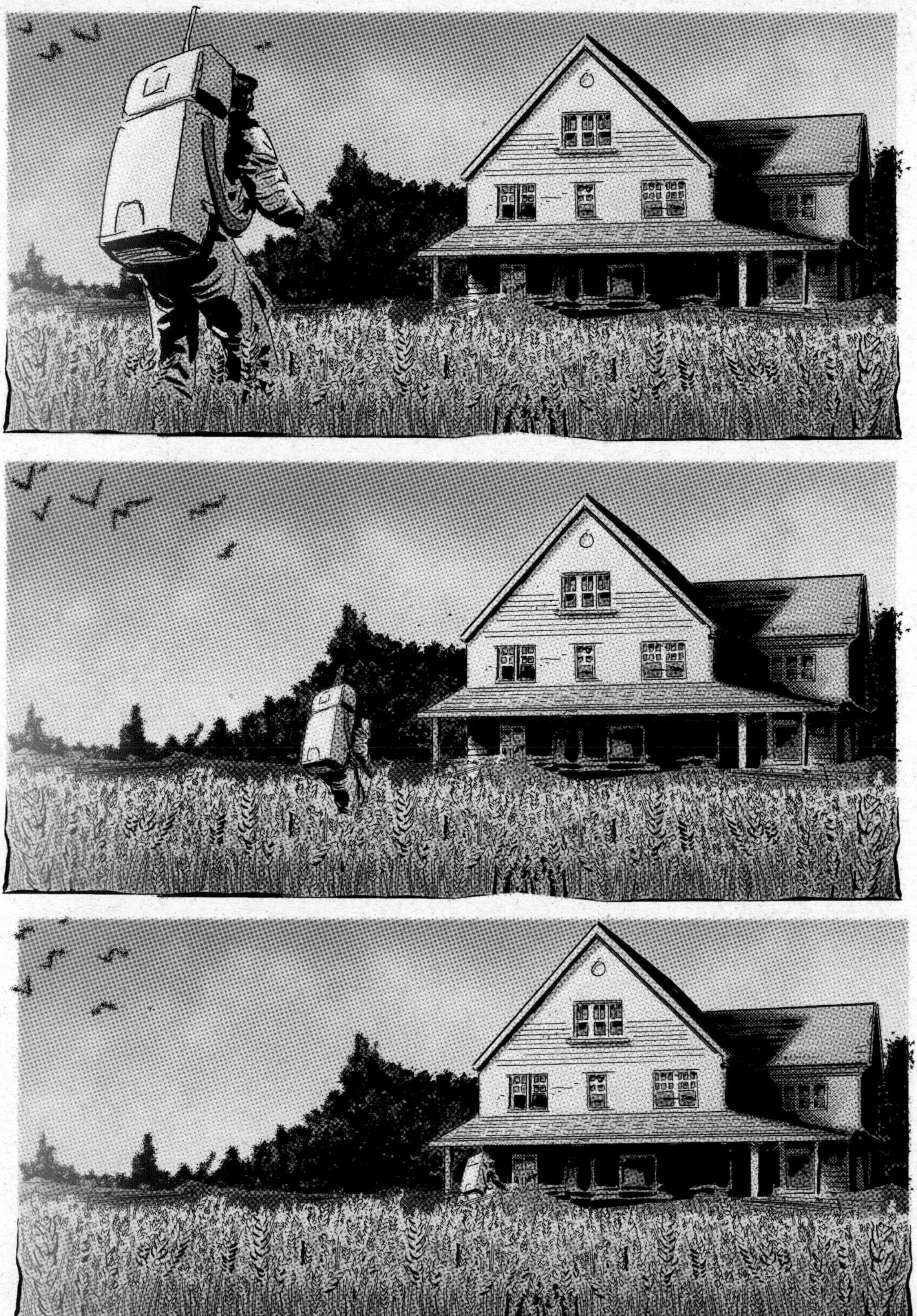

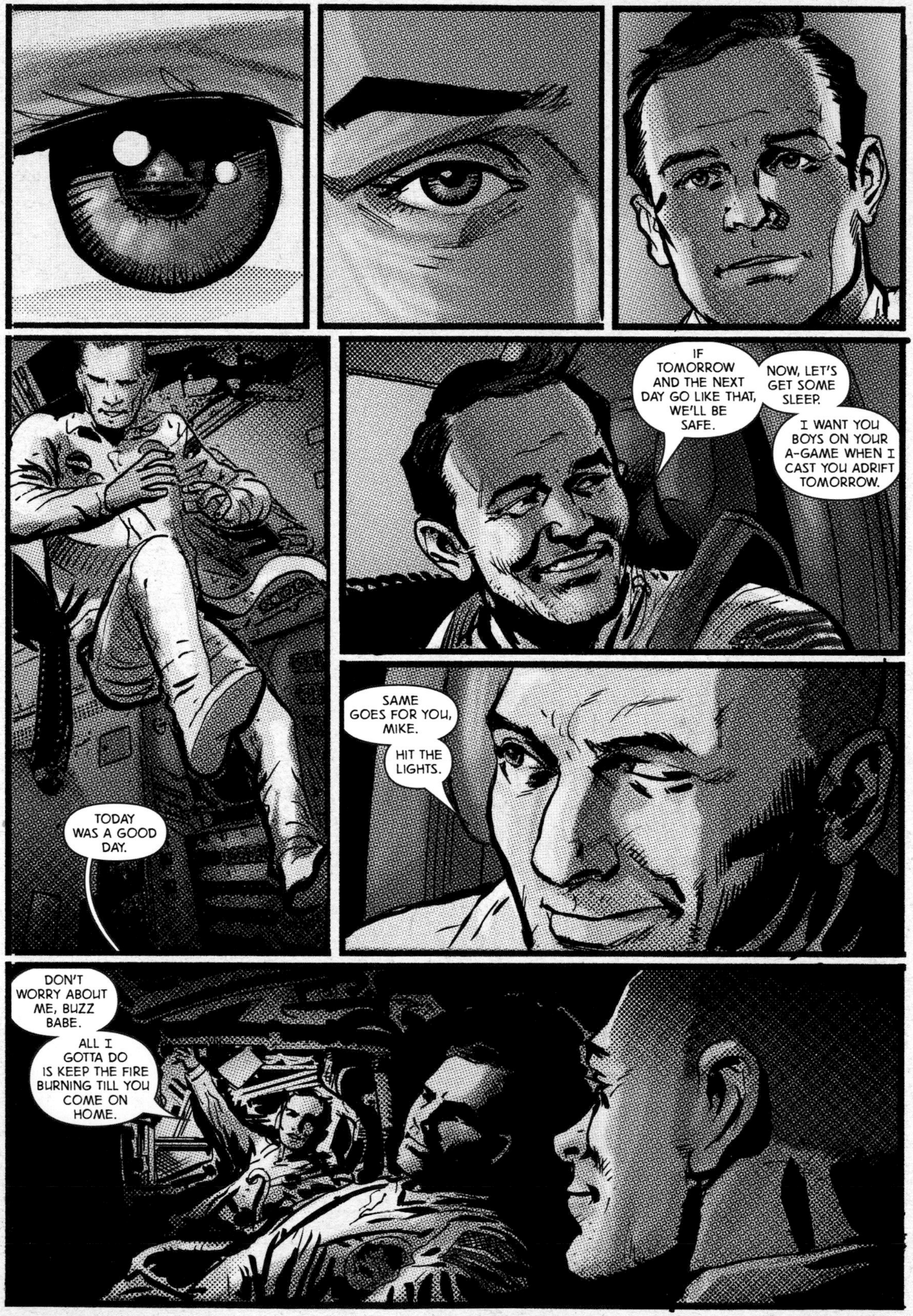
IF TOMORROW AND THE NEXT DAY GO LIKE THAT, WE'LL BE SAFE.
NOW, LET'S GET SOME SLEEP.
I WANT YOU BOYS ON YOUR A-GAME WHEN I CAST YOU ADRIFT TOMORROW.
TODAY WAS A GOOD DAY.
SAME GOES FOR YOU, MIKE.
HIT THE LIGHTS.
DON'T WORRY ABOUT ME, BUZZ BABE.
ALL I GOTTA DO IS KEEP THE FIRE BURNING TILL YOU COME ON HOME.

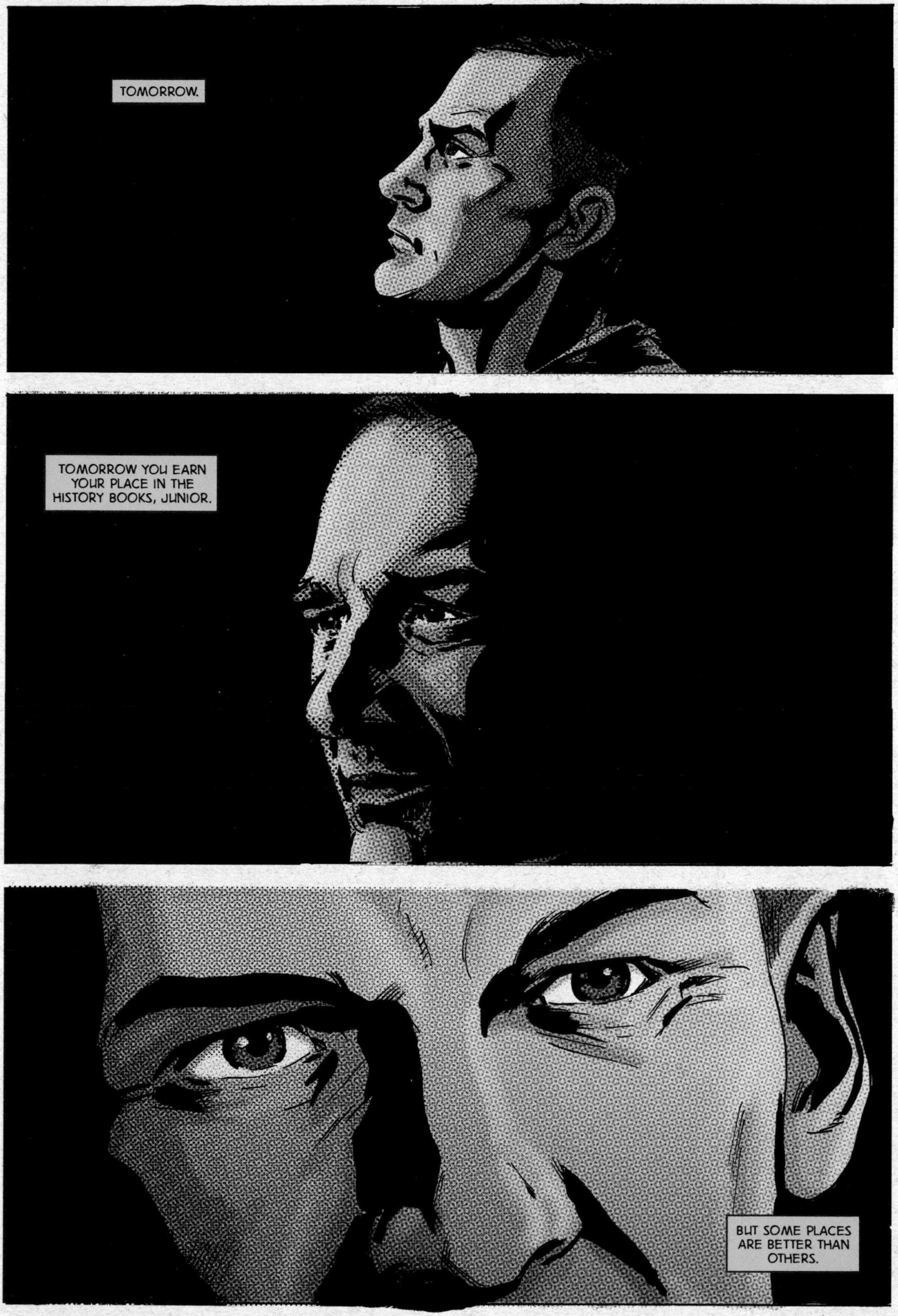
TOMORROW.
TOMORROW YOU EARN YOUR PLACE IN THE HISTORY BOOKS, JUNIOR.
BUT SOME PLACES ARE BETTER THAN OTHERS.

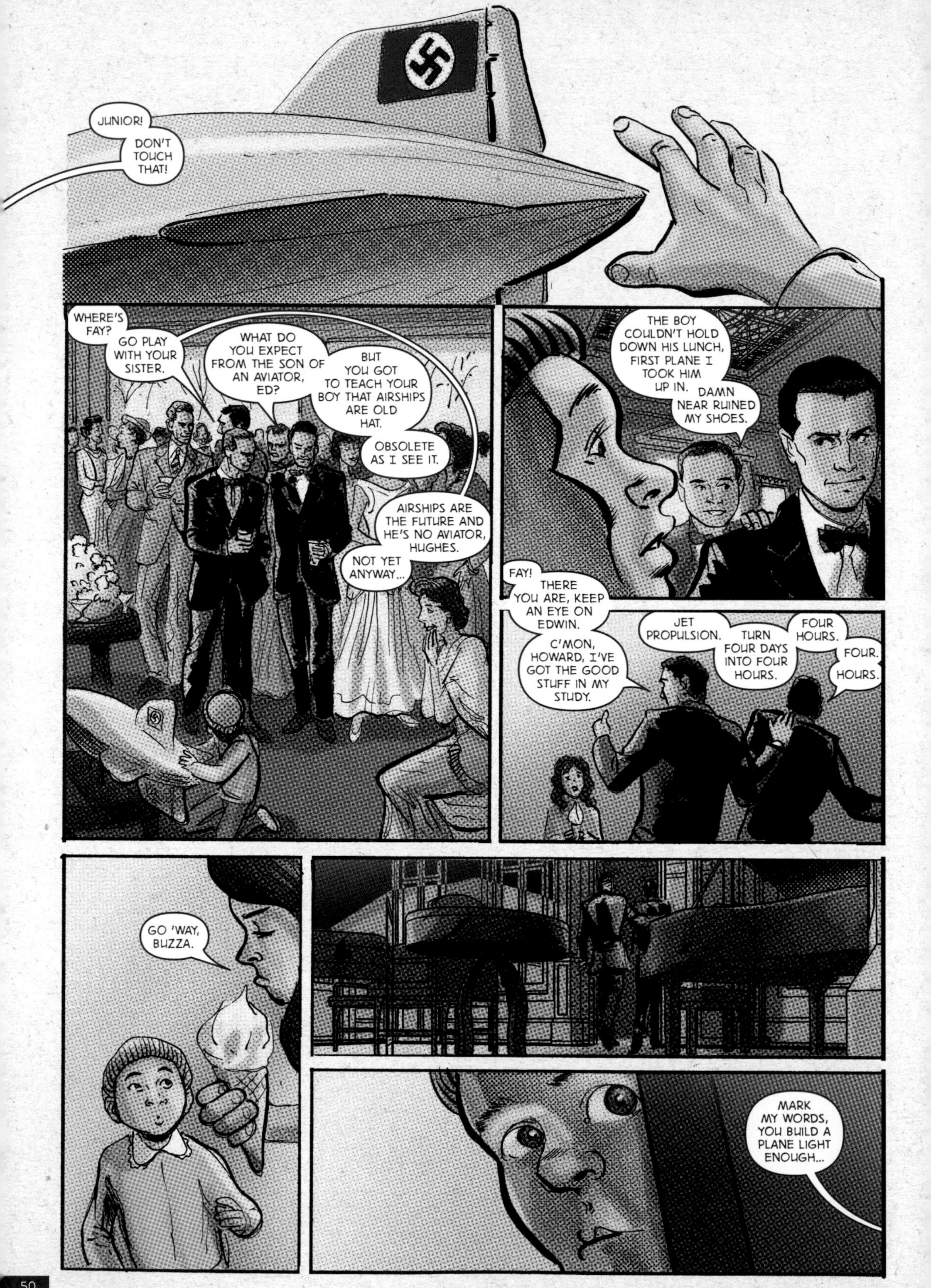
JUNIOR!
DON'T TOUCH THAT!
WHERE'S FAY?
GO PLAY WITH YOUR SISTER.
WHAT DO YOU EXPECT FROM THE SON OF AN AVIATOR, ED?
BUT YOU GOT TO TEACH YOUR BOY THAT AIRSHIPS ARE OLD HAT.
OBSOLETE AS I SEE IT.
AIRSHIPS ARE THE FUTURE AND HE'S NO AVIATOR, HUGHES.
NOT YET ANYWAY...
THE BOY COULDN'T HOLD DOWN HIS LUNCH, FIRST PLANE I TOOK HIM UP IN.
DAMN NEAR RUINED MY SHOES.
FAY!
THERE YOU ARE, KEEP AN EYE ON EDWIN.
C'MON, HOWARD, I'VE GOT THE GOOD STUFF IN MY STUDY.
JET PROPULSION.
TURN FOUR DAYS INTO FOUR HOURS.
FOUR HOURS.
FOUR.
HOURS.
GO 'WAY, BUZZA.
MARK MY WORDS, YOU BUILD A PLANE LIGHT ENOUGH...

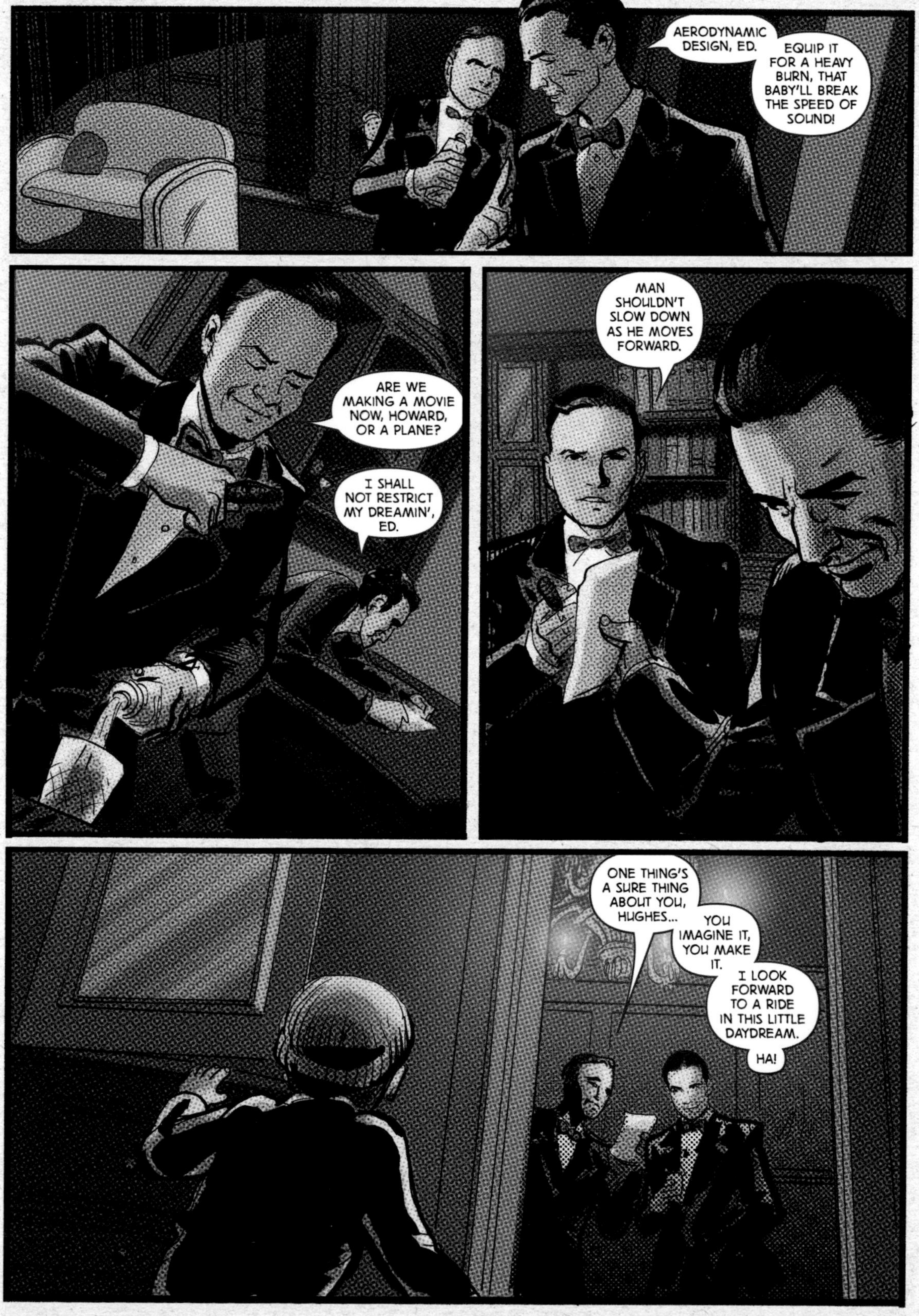
AERODYNAMIC DESIGN, ED.
EQUIP IT FOR A HEAVY BURN, THAT BABY'LL BREAK THE SPEED OF SOUND!
ARE WE MAKING A MOVIE NOW, HOWARD, OR A PLANE?
I SHALL NOT RESTRICT MY DREAMIN', ED.
MAN SHOULDN'T SLOW DOWN AS HE MOVES FORWARD.
ONE THING'S A SURE THING ABOUT YOU, HUGHES...
YOU IMAGINE IT, YOU MAKE IT.
I LOOK FORWARD TO A RIDE IN THIS LITTLE DAYDREAM.
HA!

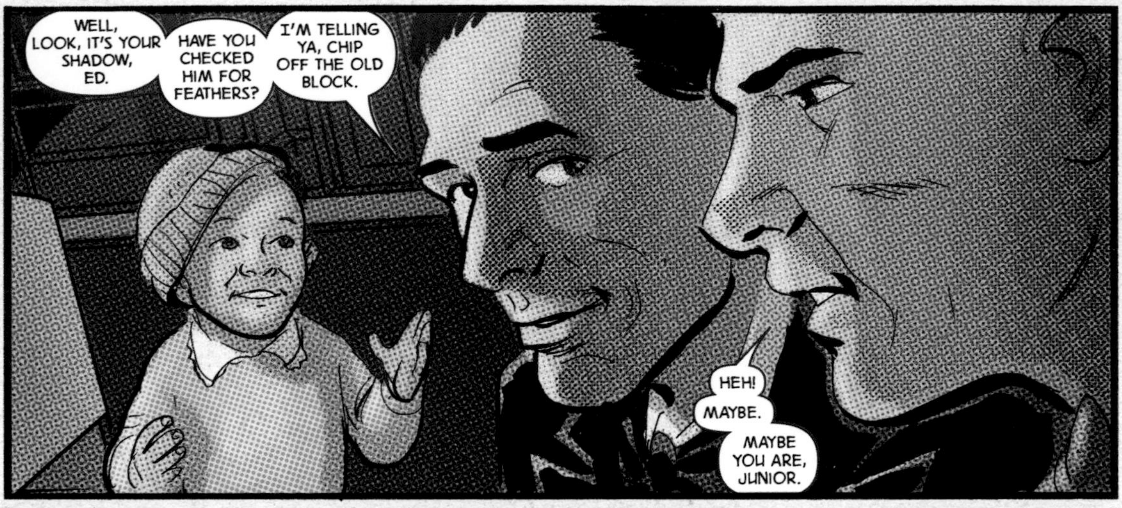
WELL, LOOK, IT'S YOUR SHADOW, ED.
HAVE YOU CHECKED HIM FOR FEATHERS?
I'M TELLING YA, CHIP OFF THE OLD BLOCK.
HEH!
MAYBE.
MAYBE YOU ARE, JUNIOR.

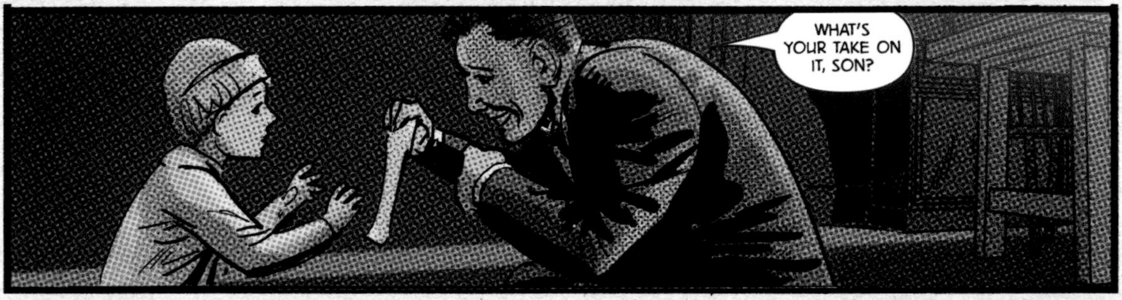
WHAT'S YOUR TAKE ON IT, SON?

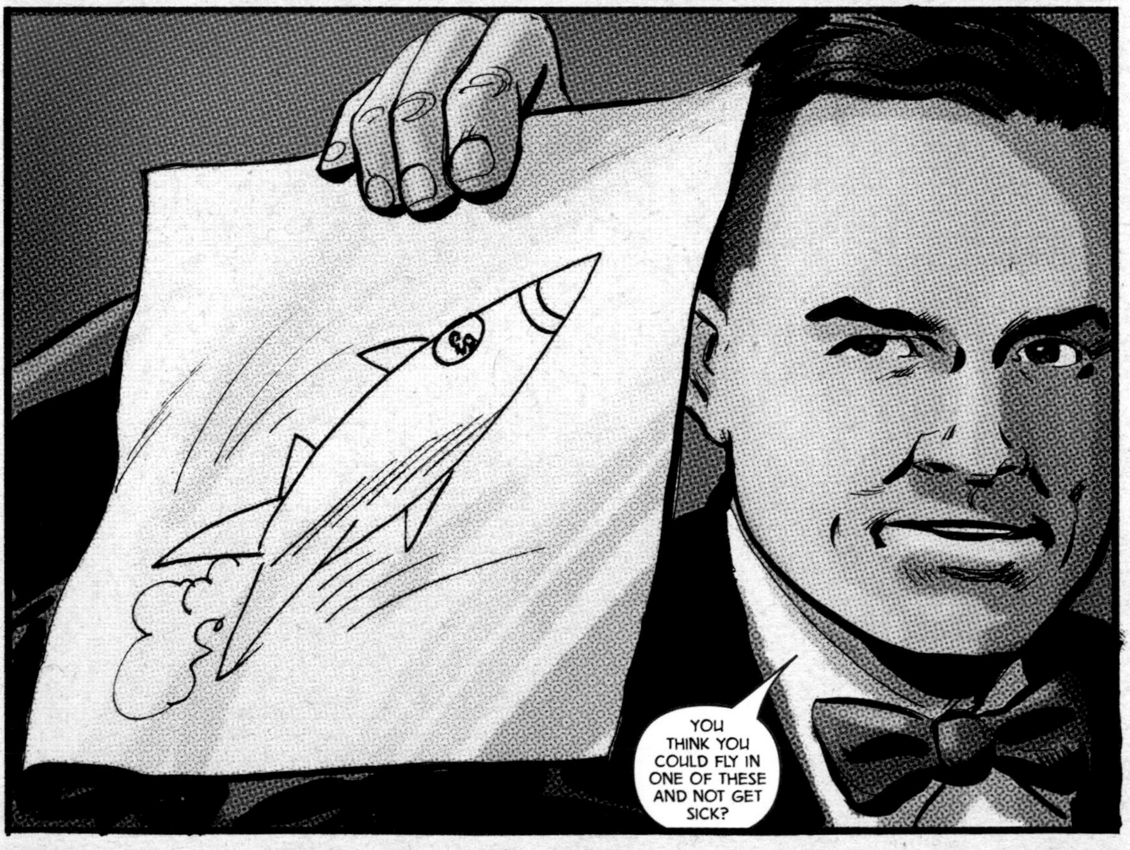
YOU THINK YOU COULD FLY IN ONE OF THESE AND NOT GET SICK?

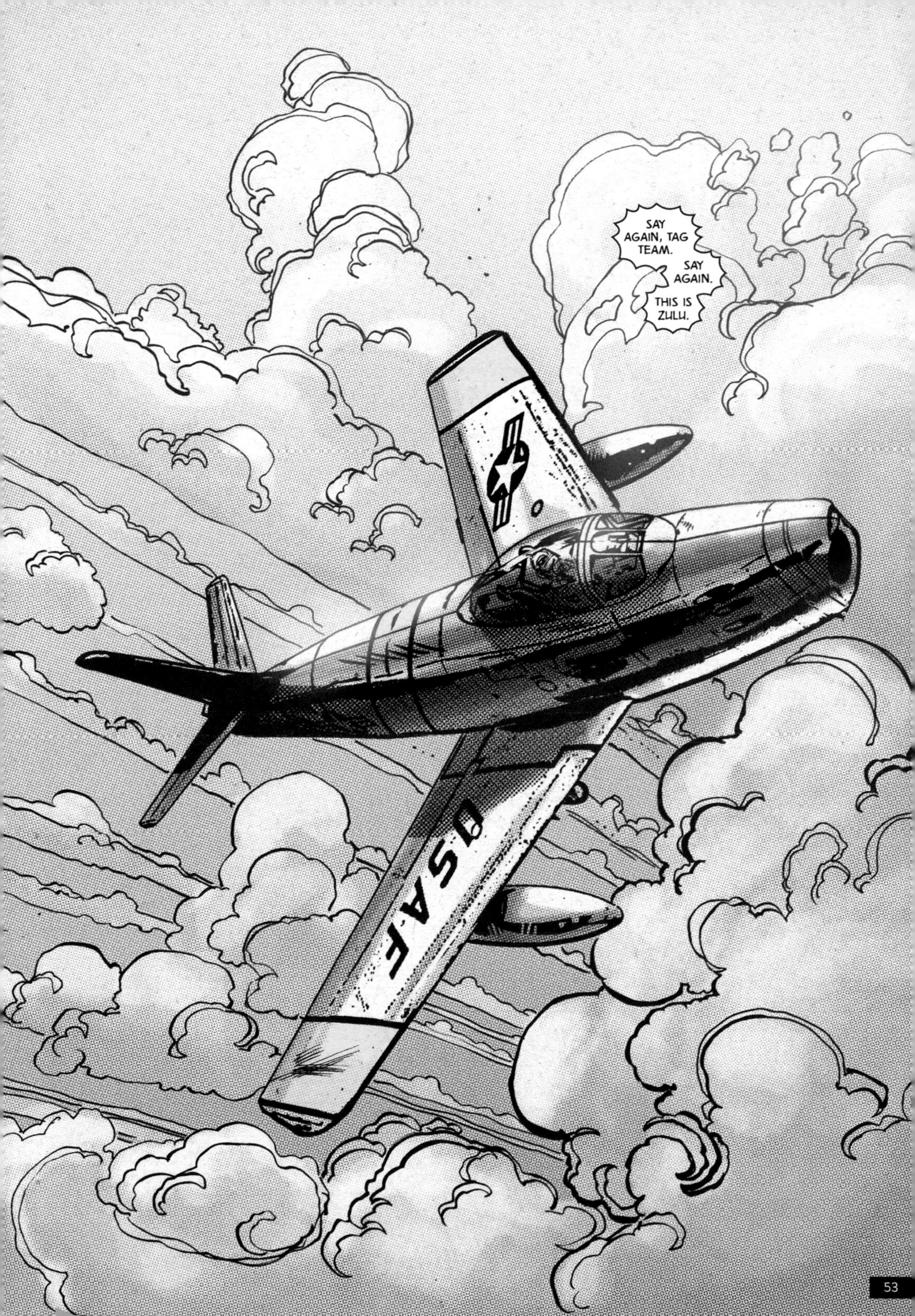
SAY AGAIN, TAG TEAM.
SAY AGAIN.
THIS IS ZULU.
USAF

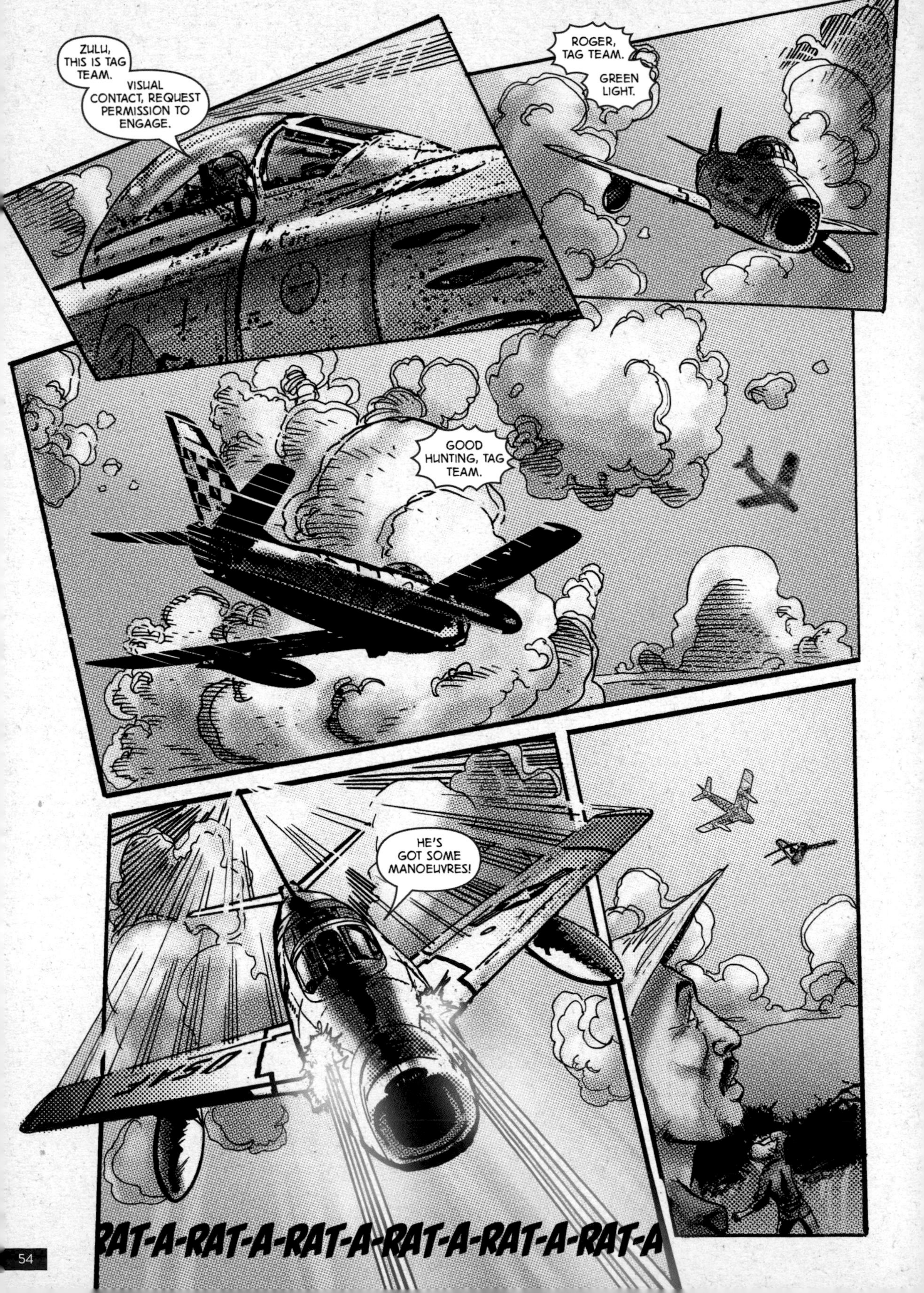
ZULU, THIS IS TAG TEAM.
VISUAL CONTACT, REQUEST PERMISSION TO ENGAGE.
ROGER, TAG TEAM.
GREEN LIGHT.
GOOD HUNTING, TAG TEAM.
HE'S GOT SOME MANOEUVRES!
RAT-A-RAT-A-RAT-A-RAT-A-RAT-A-RAT-A

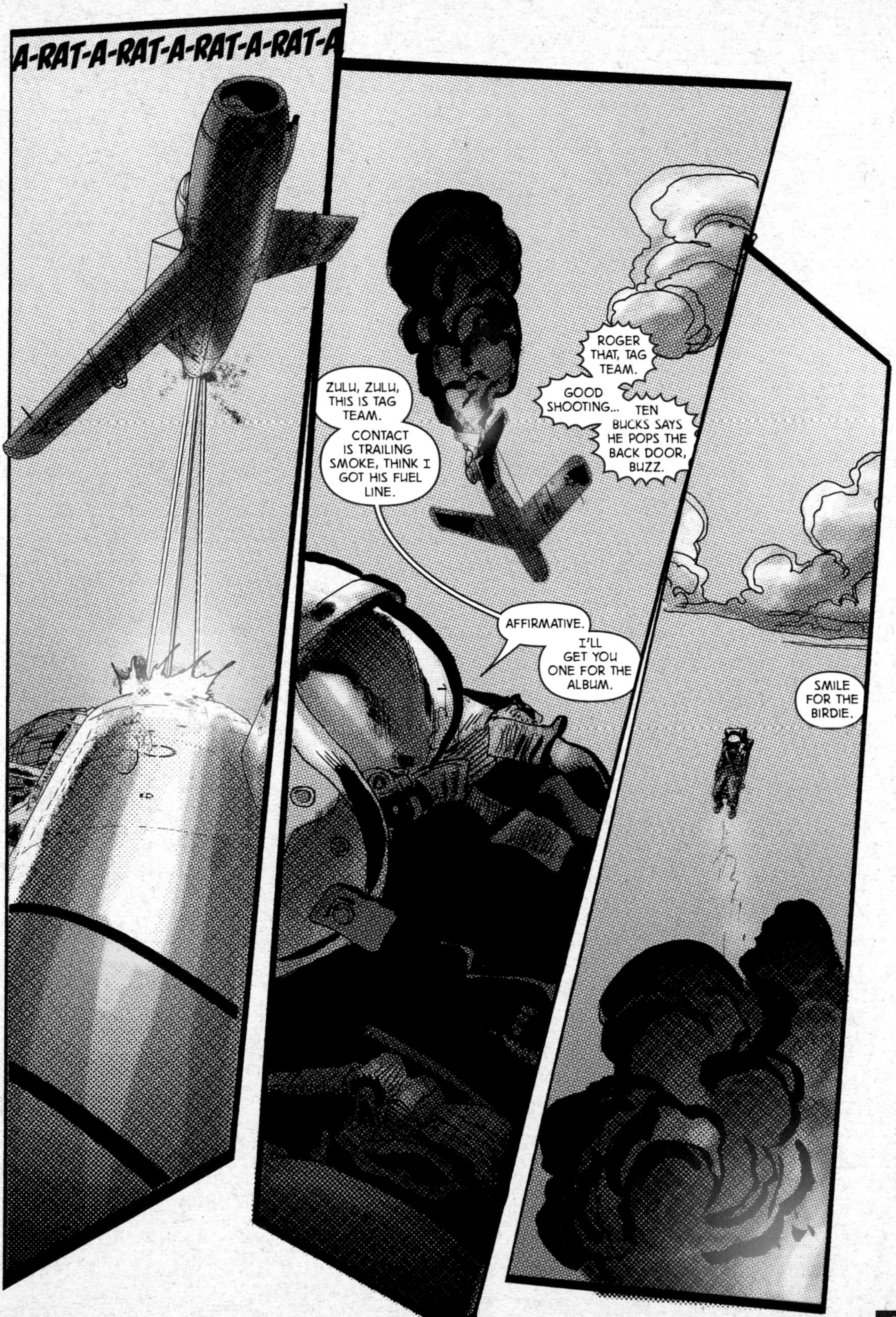
A-RAT-A-RAT-A-RAT-A-RAT-A
ZULU, ZULU, THIS IS TAG TEAM.
CONTACT IS TRAILING SMOKE, THINK I GOT HIS FUEL LINE.
ROGER THAT, TAG TEAM.
GOOD SHOOTING...
TEN BUCKS SAYS HE POPS THE BACK DOOR, BUZZ.
AFFIRMATIVE.
I'LL GET YOU ONE FOR THE ALBUM.
SMILE FOR THE BIRDIE.

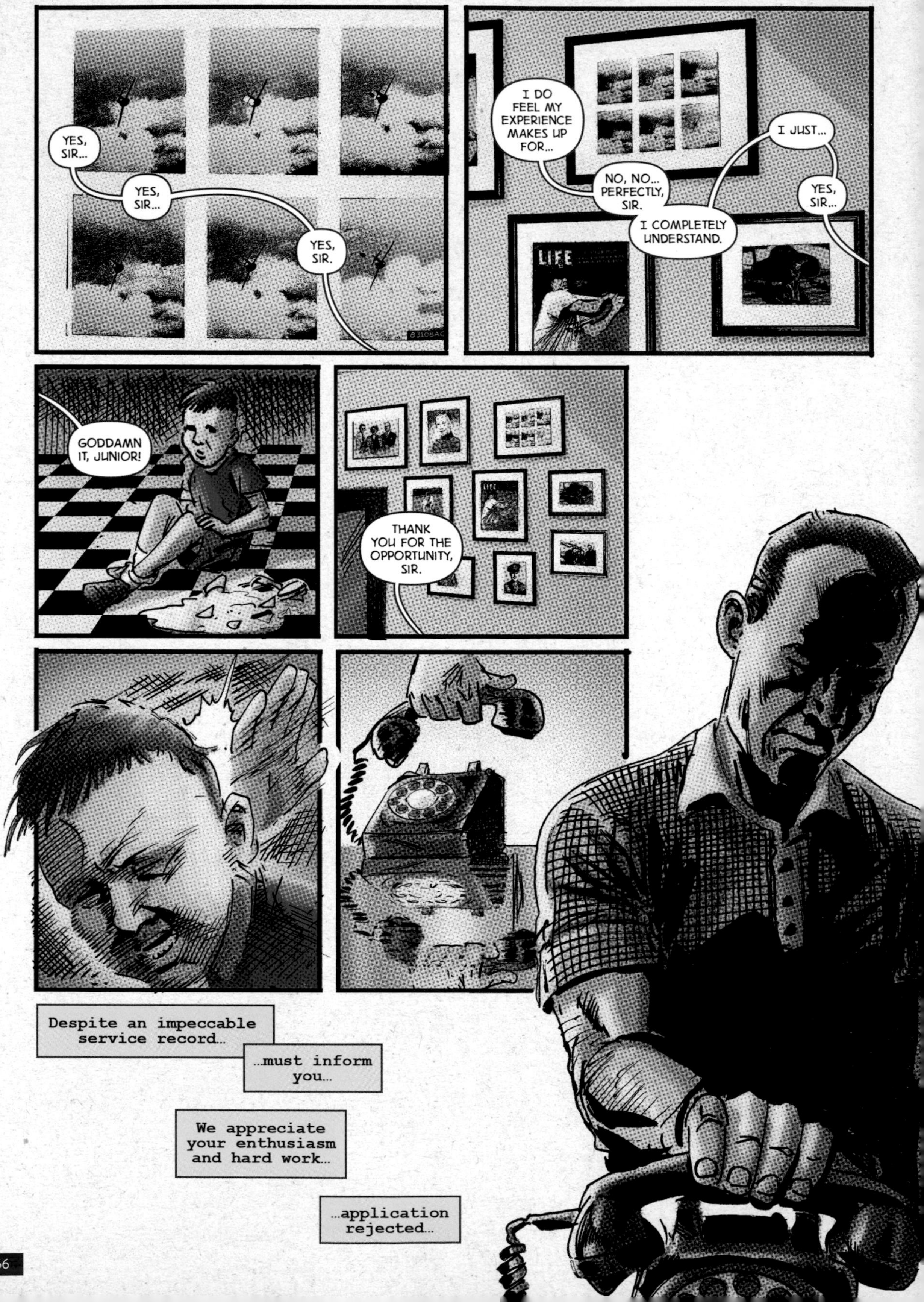

YES, SIR...
YES, SIR...
YES, SIR.
8310BAC
I DO FEEL MY EXPERIENCE MAKES UP FOR...
NO, NO... PERFECTLY, SIR.
I COMPLETELY UNDERSTAND.
I JUST...
YES, SIR...
LIFE
GODDAMN IT, JUNIOR!
LIFE
THANK YOU FOR THE OPPORTUNITY, SIR.
Despite an impeccable service record...
...must inform you...
We appreciate your enthusiasm and hard work...
...application rejected...

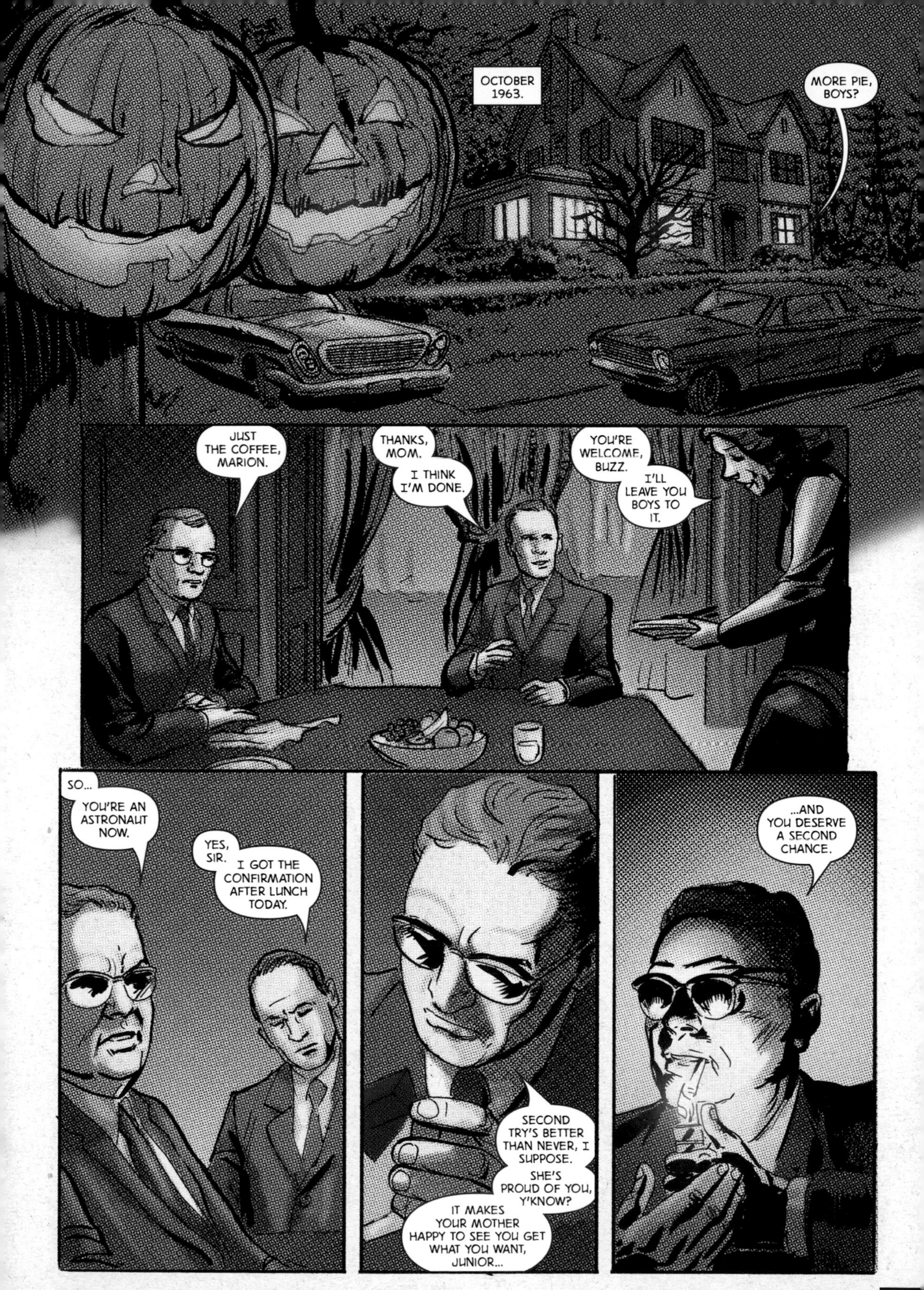

OCTOBER 1963.
MORE PIE, BOYS?
JUST THE COFFEE, MARION.
THANKS, MOM.
I THINK I'M DONE.
YOU'RE WELCOME, BUZZ.
I'LL LEAVE YOU BOYS TO IT.
SO...
YOU'RE AN ASTRONAUT NOW.
YES, SIR.
I GOT THE CONFIRMATION AFTER LUNCH TODAY.
SECOND TRY'S BETTER THAN NEVER, I SUPPOSE.
SHE'S PROUD OF YOU, Y'KNOW?
IT MAKES YOUR MOTHER HAPPY TO SEE YOU GET WHAT YOU WANT, JUNIOR...
...AND YOU DESERVE A SECOND CHANCE.

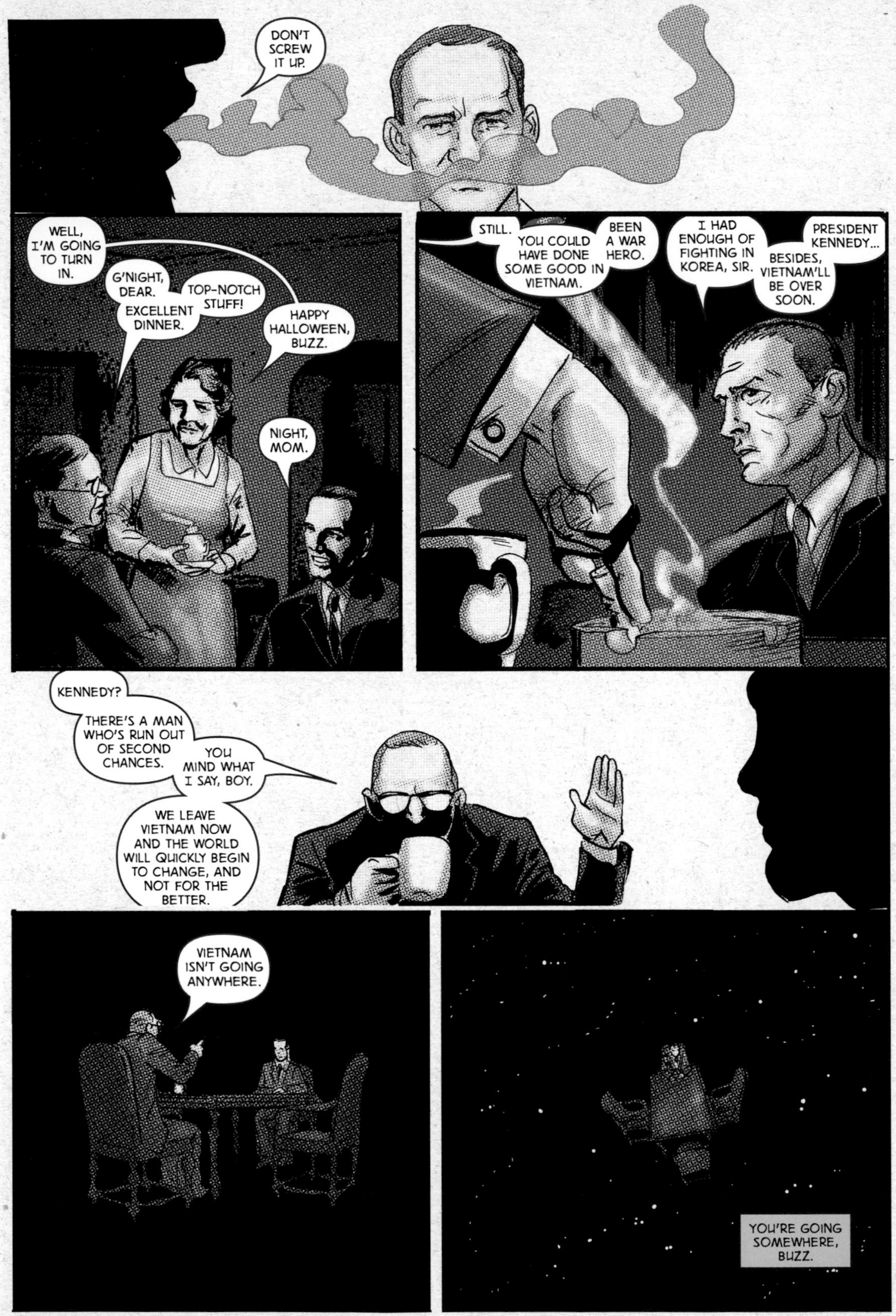
DON'T SCREW IT UP.
WELL, I'M GOING TO TURN IN.
G'NIGHT, DEAR.
EXCELLENT DINNER.
TOP-NOTCH STUFF!
HAPPY HALLOWEEN, BUZZ.
NIGHT, MOM.
STILL.
YOU COULD HAVE DONE SOME GOOD IN VIETNAM.
BEEN A WAR HERO.
I HAD ENOUGH OF FIGHTING IN KOREA, SIR.
PRESIDENT KENNEDY...
BESIDES, VIETNAM'LL BE OVER SOON.
KENNEDY?
THERE'S A MAN WHO'S RUN OUT OF SECOND CHANCES.
YOU MIND WHAT I SAY, BOY.
WE LEAVE VIETNAM NOW AND THE WORLD WILL QUICKLY BEGIN TO CHANGE, AND NOT FOR THE BETTER.
VIETNAM ISN'T GOING ANYWHERE.
YOU'RE GOING SOMEWHERE, BUZZ.

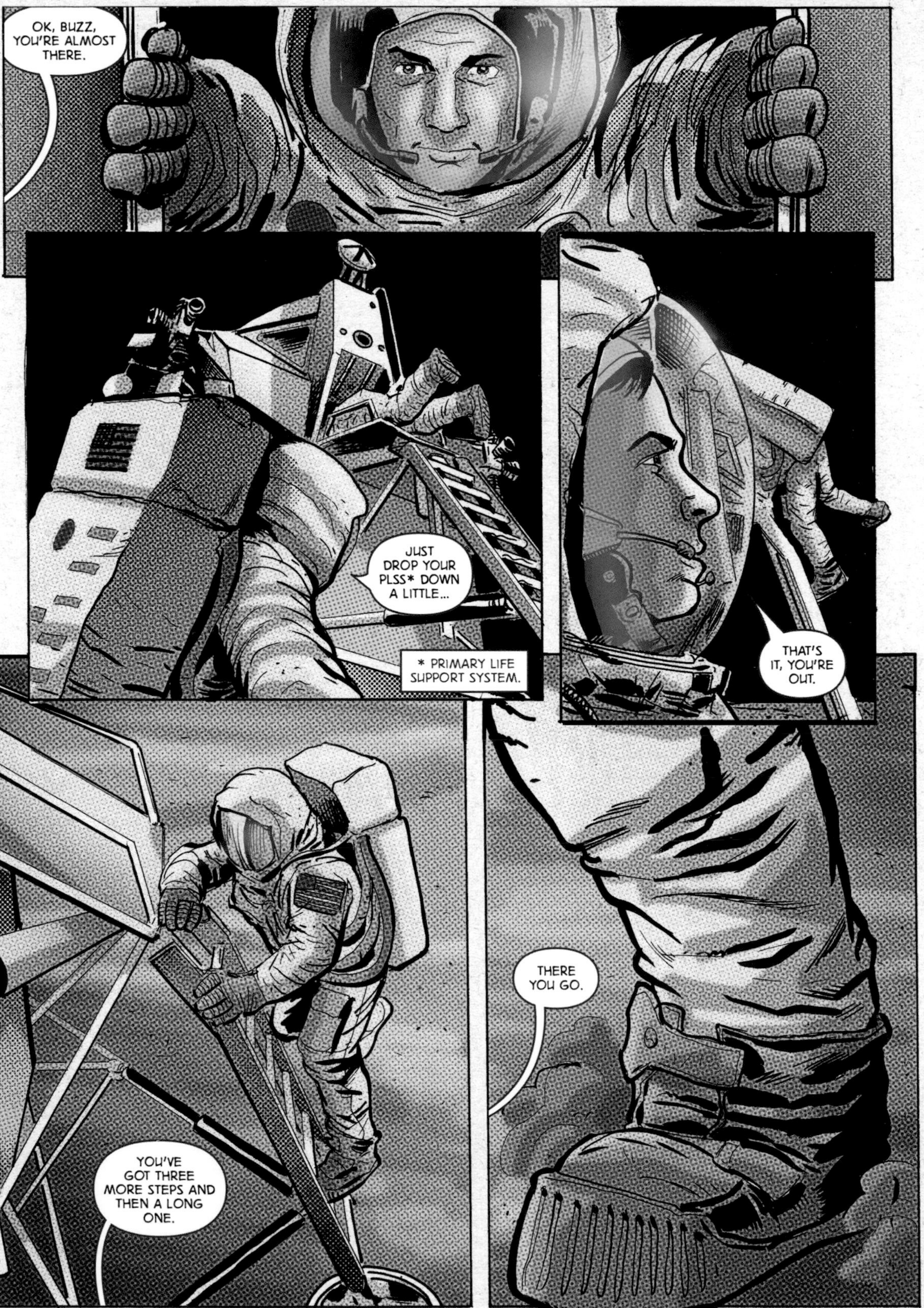
OK, BUZZ, YOU'RE ALMOST THERE.
JUST DROP YOUR PLSS* DOWN A LITTLE...
* PRIMARY LIFE SUPPORT SYSTEM.
THAT'S IT, YOU'RE OUT.
YOU'VE GOT THREE MORE STEPS AND THEN A LONG ONE.
THERE YOU GO.

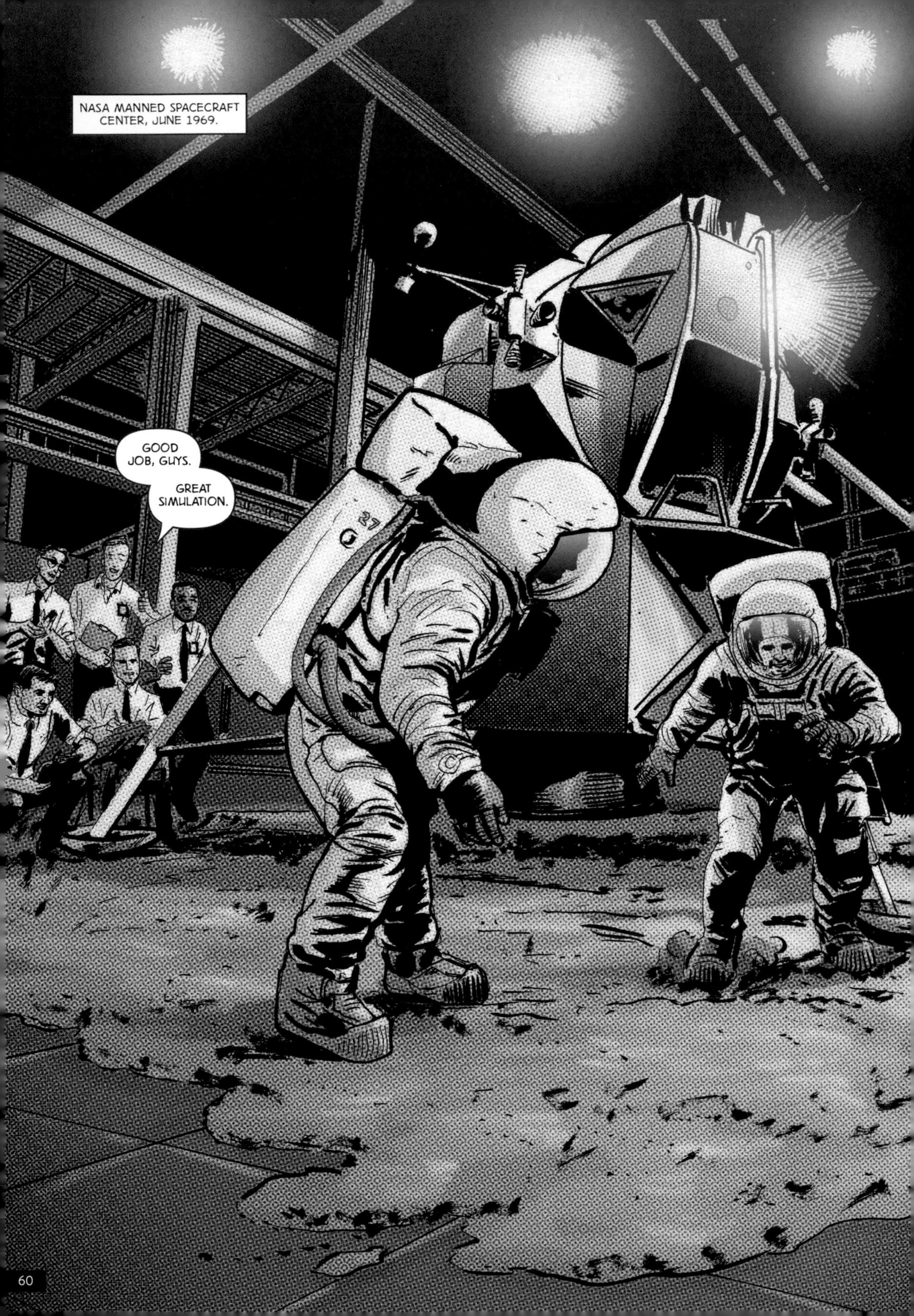
NASA MANNED SPACECRAFT CENTER, JUNE 1969.
GOOD JOB, GUYS.
GREAT SIMULATION.

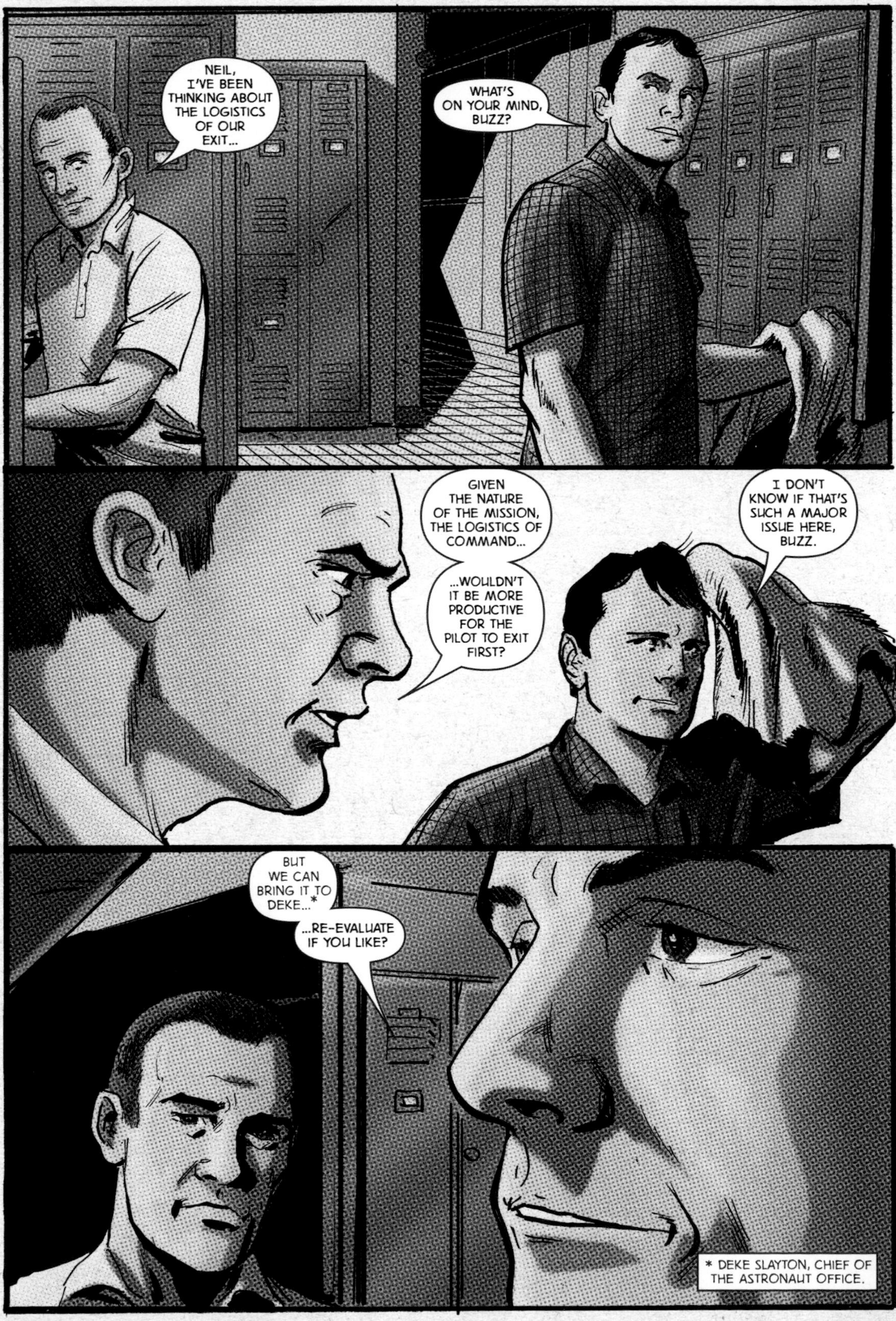
NEIL, I'VE BEEN THINKING ABOUT THE LOGISTICS OF OUR EXIT...
WHAT'S ON YOUR MIND, BUZZ?
GIVEN THE NATURE OF THE MISSION, THE LOGISTICS OF COMMAND...
...WOULDN'T IT BE MORE PRODUCTIVE FOR THE PILOT TO EXIT FIRST?
I DON'T KNOW IF THAT'S SUCH A MAJOR ISSUE HERE, BUZZ.
BUT WE CAN BRING IT TO DEKE...*
...RE-EVALUATE IF YOU LIKE?
* DEKE SLAYTON, CHIEF OF THE ASTRONAUT OFFICE.

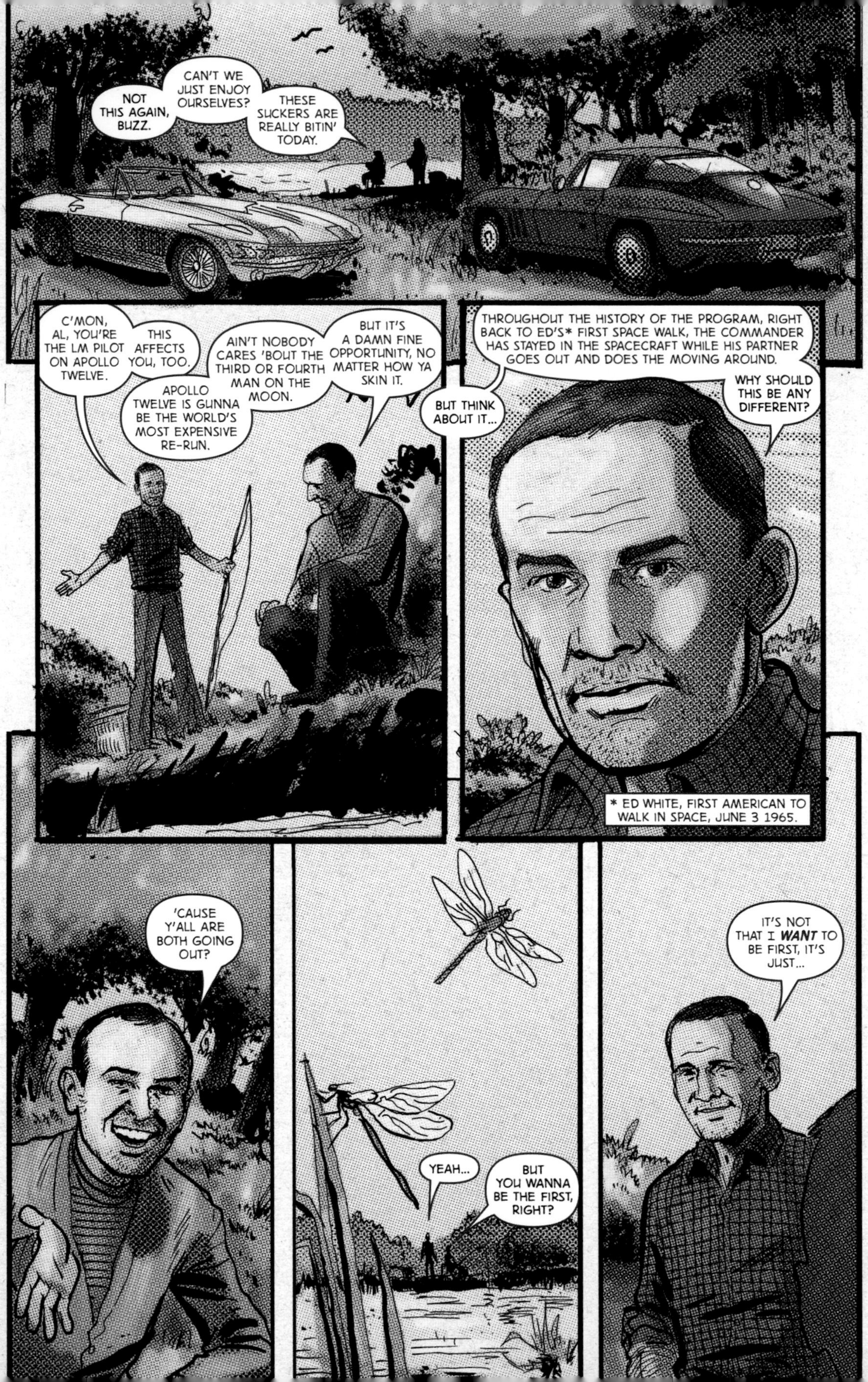
NOT THIS AGAIN, BUZZ.
CAN'T WE JUST ENJOY OURSELVES?
THESE SUCKERS ARE REALLY BITIN' TODAY.
C'MON, AL, YOU'RE THE LM PILOT ON APOLLO TWELVE.
THIS AFFECTS YOU, TOO.
APOLLO TWELVE IS GUNNA BE THE WORLD'S MOST EXPENSIVE RE-RUN.
AIN'T NOBODY CARES 'BOUT THE THIRD OR FOURTH MAN ON THE MOON.
BUT IT'S A DAMN FINE OPPORTUNITY, NO MATTER HOW YA SKIN IT.
BUT THINK ABOUT IT...
THROUGHOUT THE HISTORY OF THE PROGRAM, RIGHT BACK TO ED'S* FIRST SPACE WALK, THE COMMANDER HAS STAYED IN THE SPACECRAFT WHILE HIS PARTNER GOES OUT AND DOES THE MOVING AROUND.
WHY SHOULD THIS BE ANY DIFFERENT?
* ED WHITE, FIRST AMERICAN TO WALK IN SPACE, JUNE 3 1965.
'CAUSE Y'ALL ARE BOTH GOING OUT?
YEAH...
BUT YOU WANNA BE THE FIRST, RIGHT?
IT'S NOT THAT I ***WANT*** TO BE FIRST, IT'S JUST...

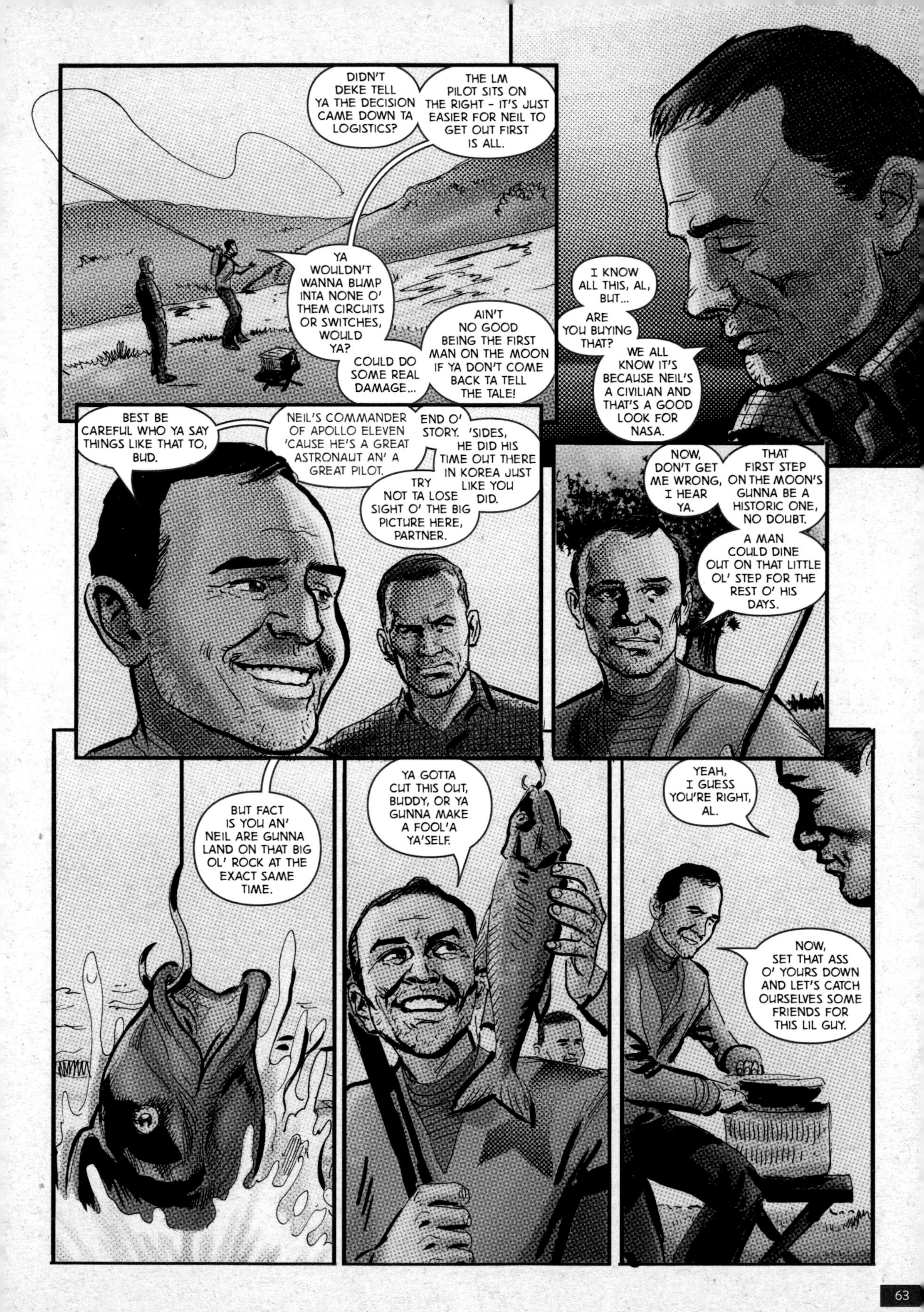
DIDN'T DEKE TELL YA THE DECISION CAME DOWN TA LOGISTICS?
THE LM PILOT SITS ON THE RIGHT - IT'S JUST EASIER FOR NEIL TO GET OUT FIRST IS ALL.
YA WOULDN'T WANNA BUMP INTA NONE O' THEM CIRCUITS OR SWITCHES, WOULD YA?
COULD DO SOME REAL DAMAGE...
AIN'T NO GOOD BEING THE FIRST MAN ON THE MOON IF YA DON'T COME BACK TA TELL THE TALE!
I KNOW ALL THIS, AL, BUT...
ARE YOU BUYING THAT?
WE ALL KNOW IT'S BECAUSE NEIL'S A CIVILIAN AND THAT'S A GOOD LOOK FOR NASA.
BEST BE CAREFUL WHO YA SAY THINGS LIKE THAT TO, BUD.
NEIL'S COMMANDER OF APOLLO ELEVEN 'CAUSE HE'S A GREAT ASTRONAUT AN' A GREAT PILOT.
END O' STORY. 'SIDES, HE DID HIS TIME OUT THERE IN KOREA JUST LIKE YOU DID.
TRY NOT TA LOSE SIGHT O' THE BIG PICTURE HERE, PARTNER.
NOW, DON'T GET ME WRONG, I HEAR YA.
THAT FIRST STEP ON THE MOON'S GUNNA BE A HISTORIC ONE, NO DOUBT.
A MAN COULD DINE OUT ON THAT LITTLE OL' STEP FOR THE REST O' HIS DAYS.
BUT FACT IS YOU AN' NEIL ARE GUNNA LAND ON THAT BIG OL' ROCK AT THE EXACT SAME TIME.
YA GOTTA CUT THIS OUT, BUDDY, OR YA GUNNA MAKE A FOOL'A YA'SELF.
YEAH, I GUESS YOU'RE RIGHT, AL.
NOW, SET THAT ASS O' YOURS DOWN AND LET'S CATCH OURSELVES SOME FRIENDS FOR THIS LIL GUY.

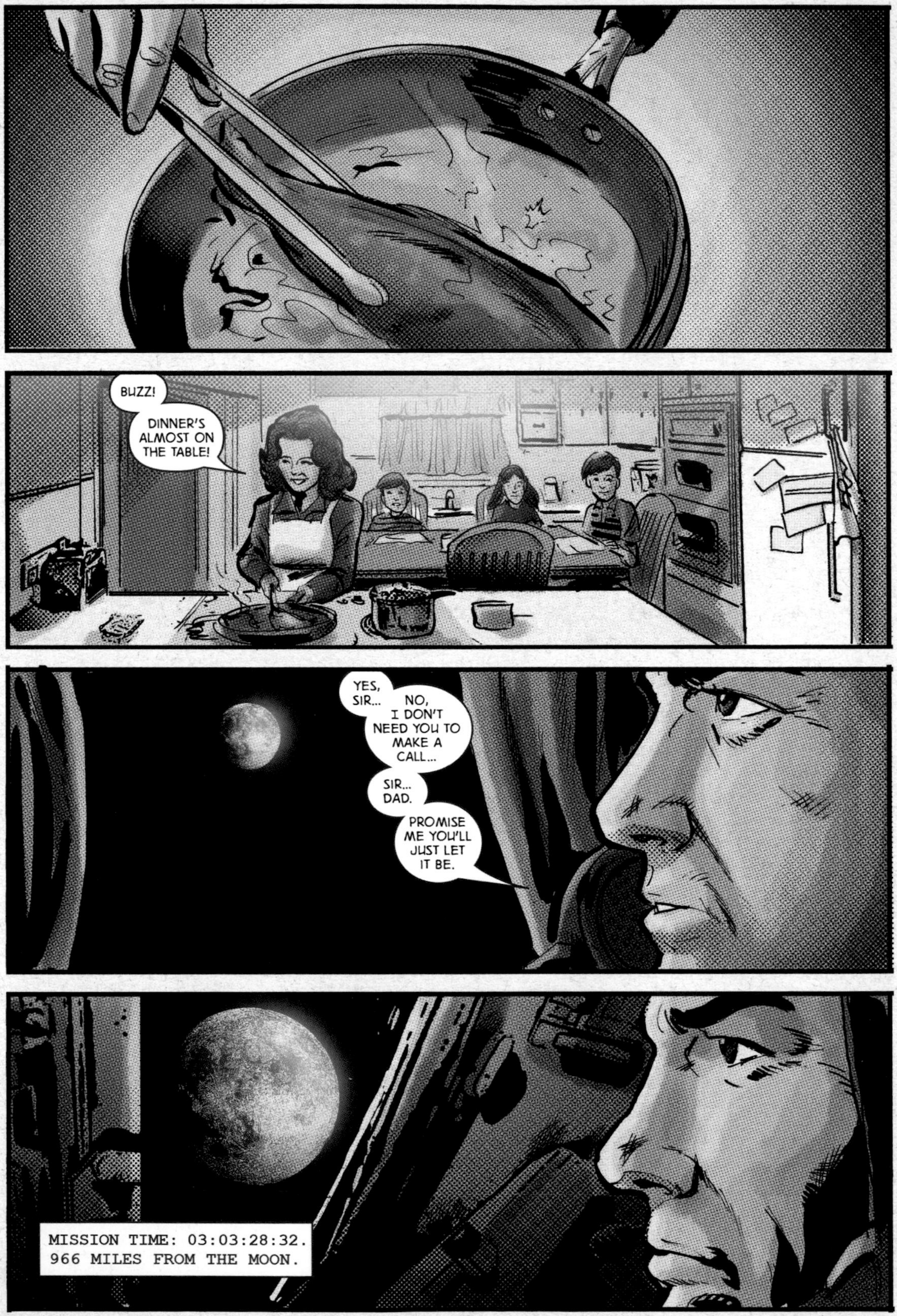

BUZZ!
DINNER'S ALMOST ON THE TABLE!
YES, SIR...
NO, I DON'T NEED YOU TO MAKE A CALL...
SIR... DAD.
PROMISE ME YOU'LL JUST LET IT BE.
MISSION TIME: 03:03:28:32.
966 MILES FROM THE MOON.

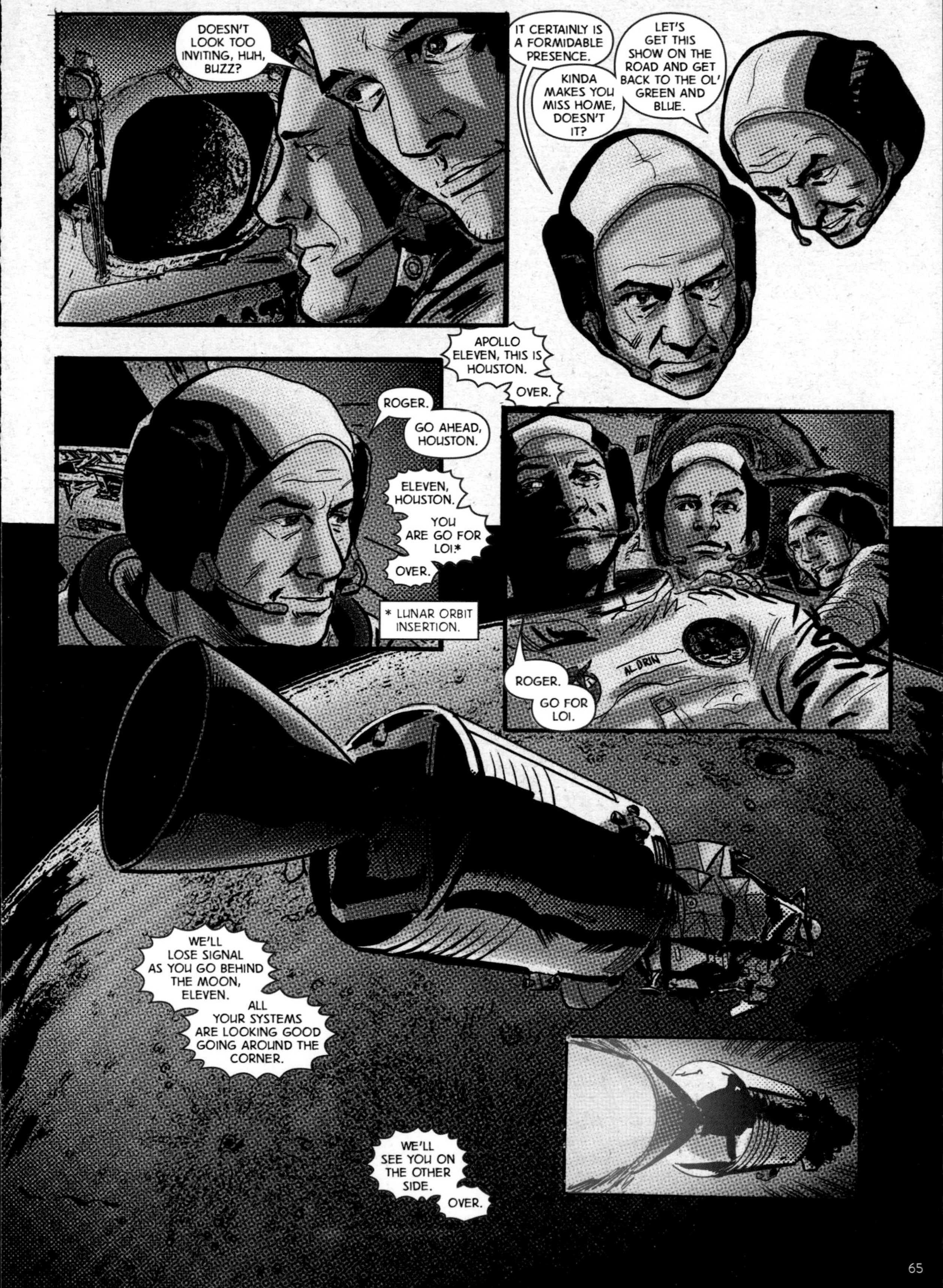
DOESN'T LOOK TOO INVITING, HUH, BUZZ?
IT CERTAINLY IS A FORMIDABLE PRESENCE.
KINDA MAKES YOU MISS HOME, DOESN'T IT?
LET'S GET THIS SHOW ON THE ROAD AND GET BACK TO THE OL' GREEN AND BLUE.
APOLLO ELEVEN, THIS IS HOUSTON.
OVER.
ROGER.
GO AHEAD, HOUSTON.
ELEVEN, HOUSTON.
YOU ARE GO FOR LOI!*
OVER.
* LUNAR ORBIT INSERTION.
ALDRIN
ROGER.
GO FOR LOI.
WE'LL LOSE SIGNAL AS YOU GO BEHIND THE MOON, ELEVEN.
ALL YOUR SYSTEMS ARE LOOKING GOOD GOING AROUND THE CORNER.
WE'LL SEE YOU ON THE OTHER SIDE.
OVER.

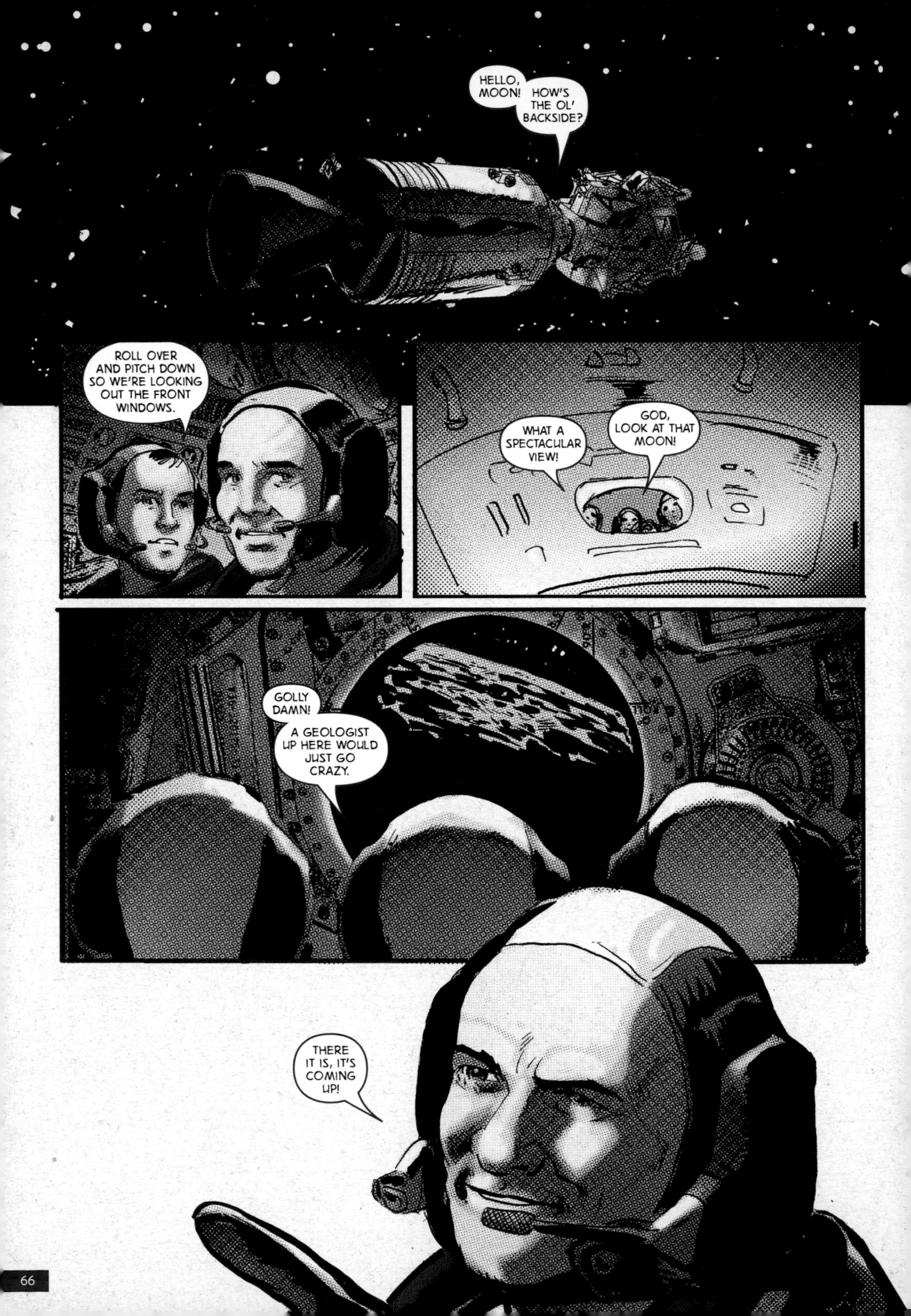
HELLO, MOON!
HOW'S THE OL' BACKSIDE?
ROLL OVER AND PITCH DOWN SO WE'RE LOOKING OUT THE FRONT WINDOWS.
WHAT A SPECTACULAR VIEW!
GOD, LOOK AT THAT MOON!
GOLLY DAMN!
A GEOLOGIST UP HERE WOULD JUST GO CRAZY.
THERE IT IS, IT'S COMING UP!

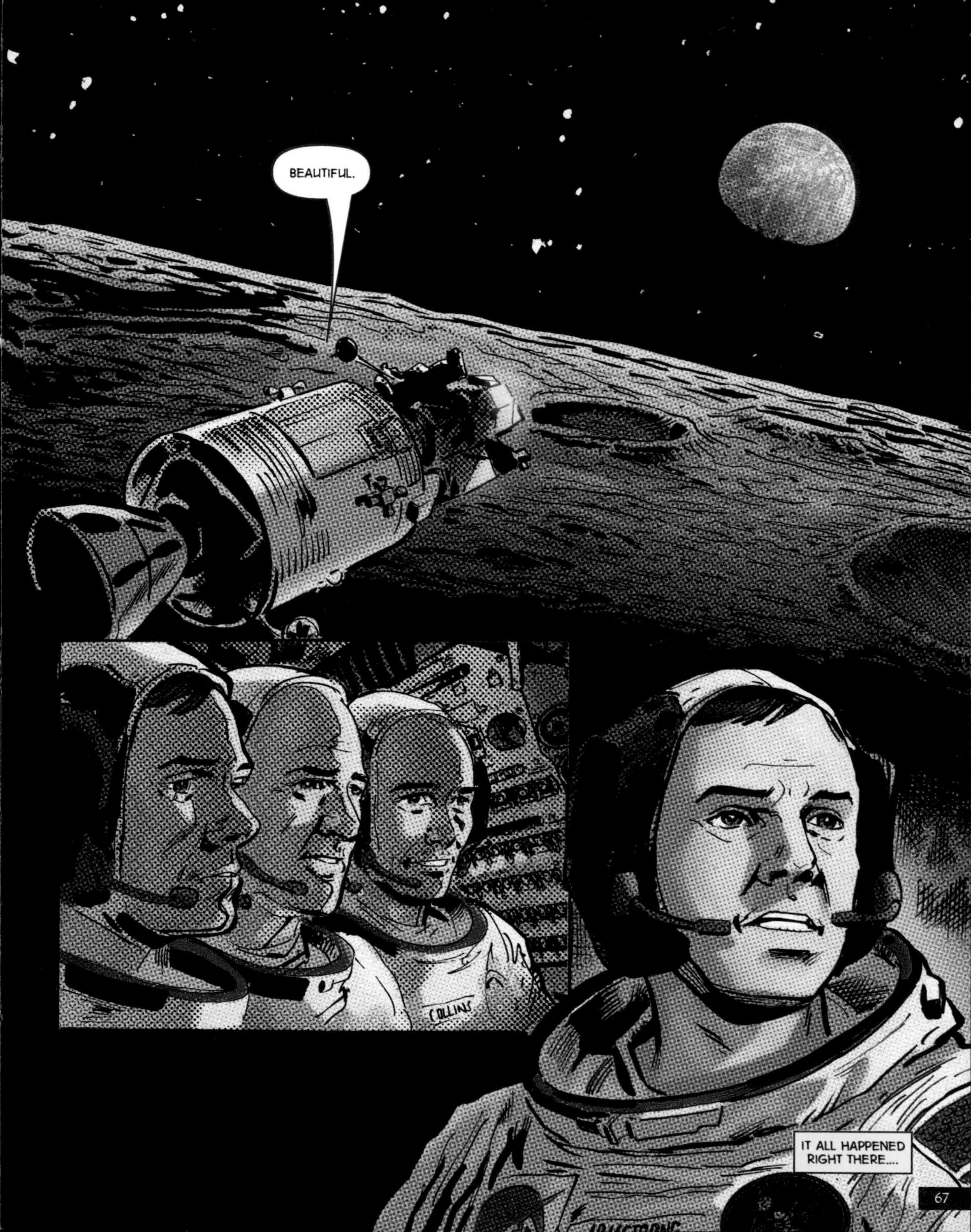
BEAUTIFUL.
COLLINS
IT ALL HAPPENED RIGHT THERE....

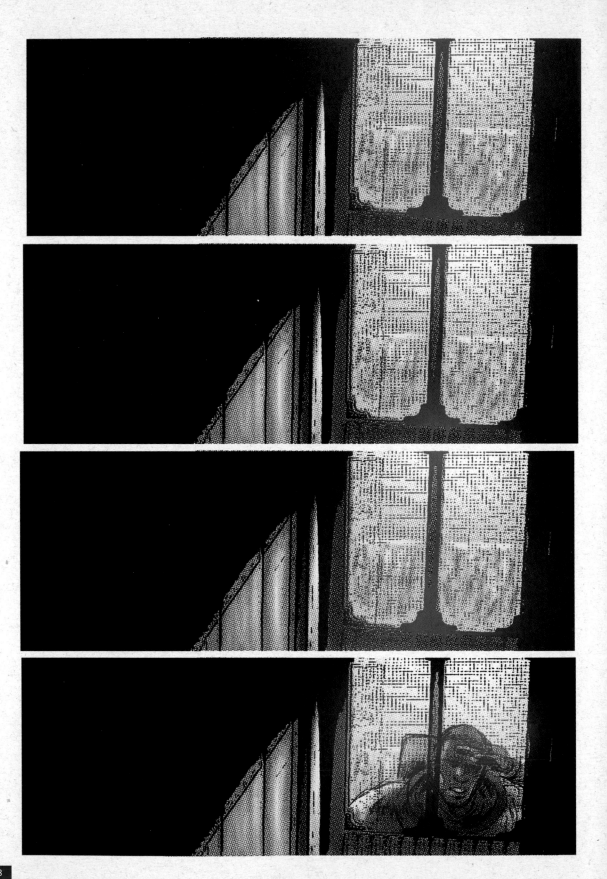

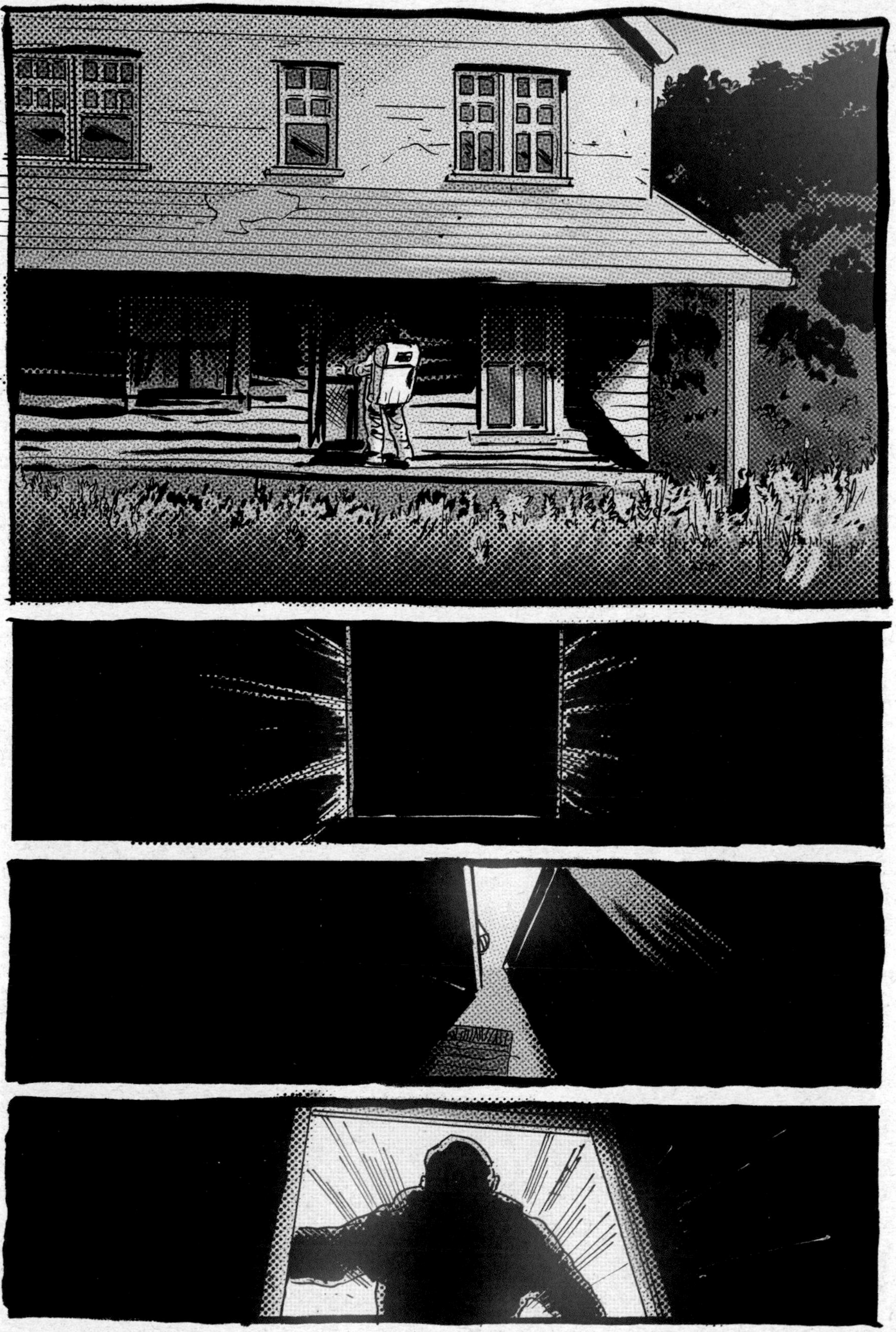

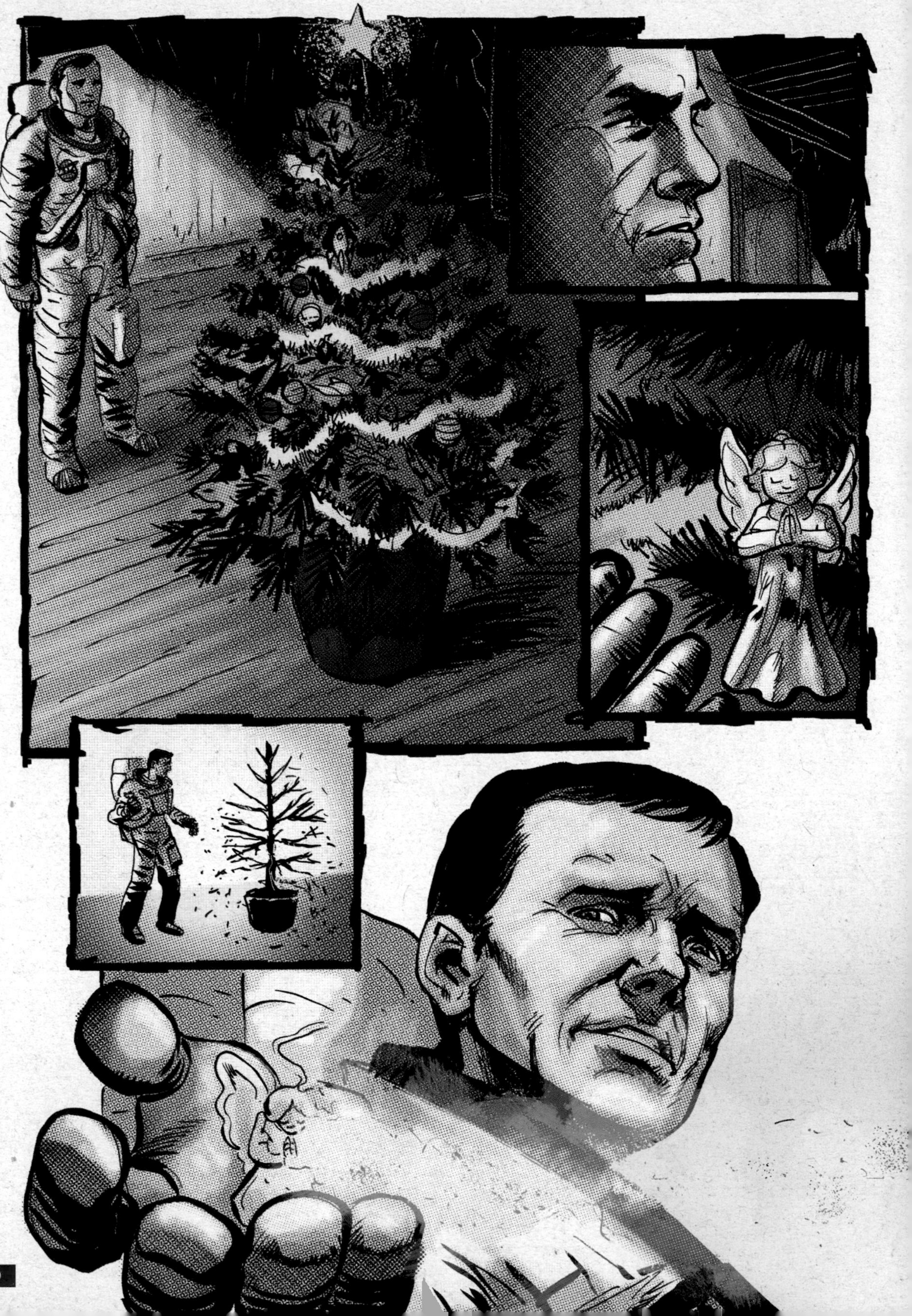

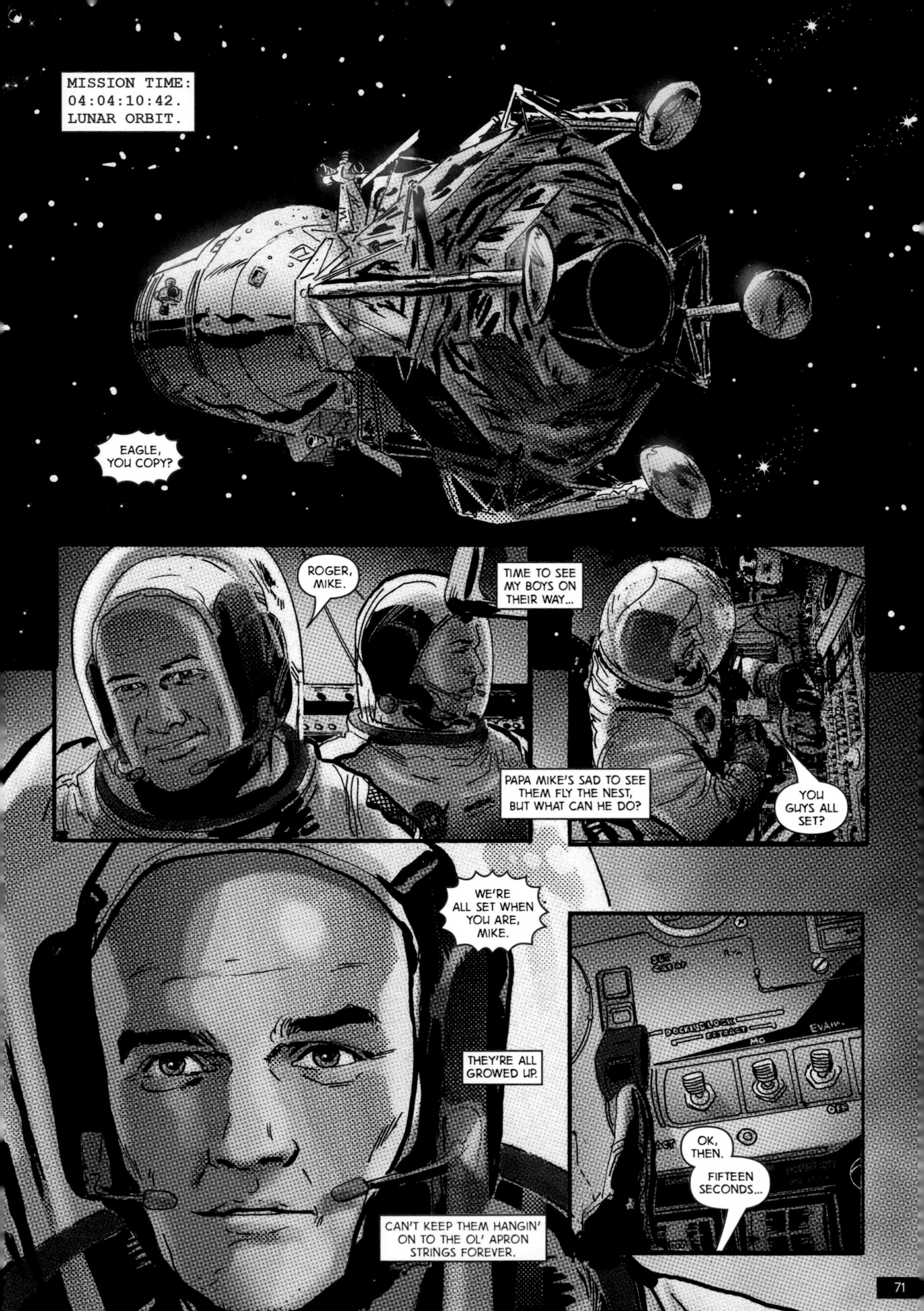
MISSION TIME: 04:04:10:42. LUNAR ORBIT.
EAGLE, YOU COPY?
ROGER, MIKE.
TIME TO SEE MY BOYS ON THEIR WAY...
PAPA MIKE'S SAD TO SEE THEM FLY THE NEST, BUT WHAT CAN HE DO?
YOU GUYS ALL SET?
WE'RE ALL SET WHEN YOU ARE, MIKE.
THEY'RE ALL GROWED UP.
CAN'T KEEP THEM HANGIN' ON TO THE OL' APRON STRINGS FOREVER.
DOCKING LOCK
RETRACT
OK, THEN.
FIFTEEN SECONDS...

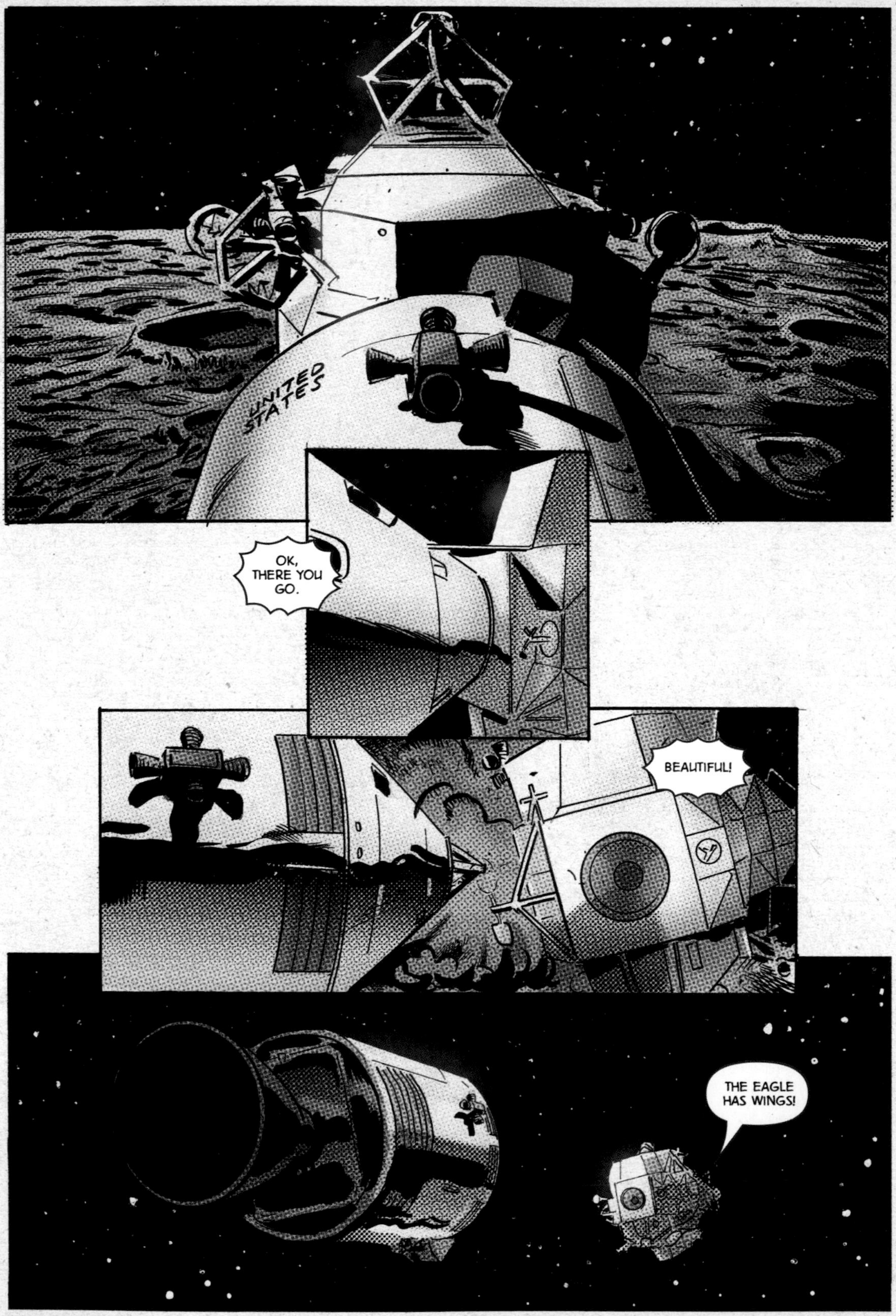
UNITED STATES
OK, THERE YOU GO.
BEAUTIFUL!
THE EAGLE HAS WINGS!

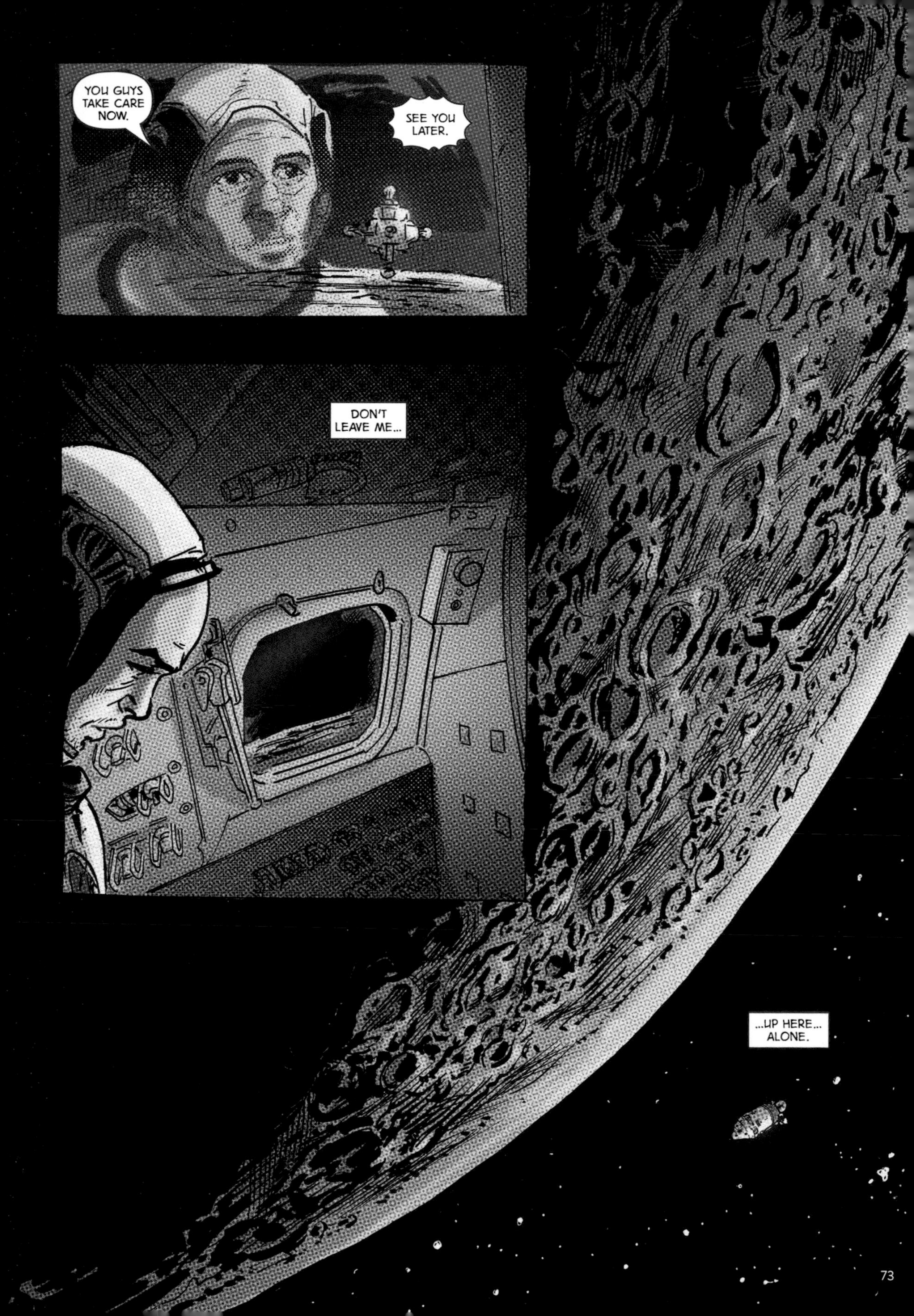
YOU GUYS TAKE CARE NOW.
SEE YOU LATER.
DON'T LEAVE ME...
...UP HERE... ALONE.

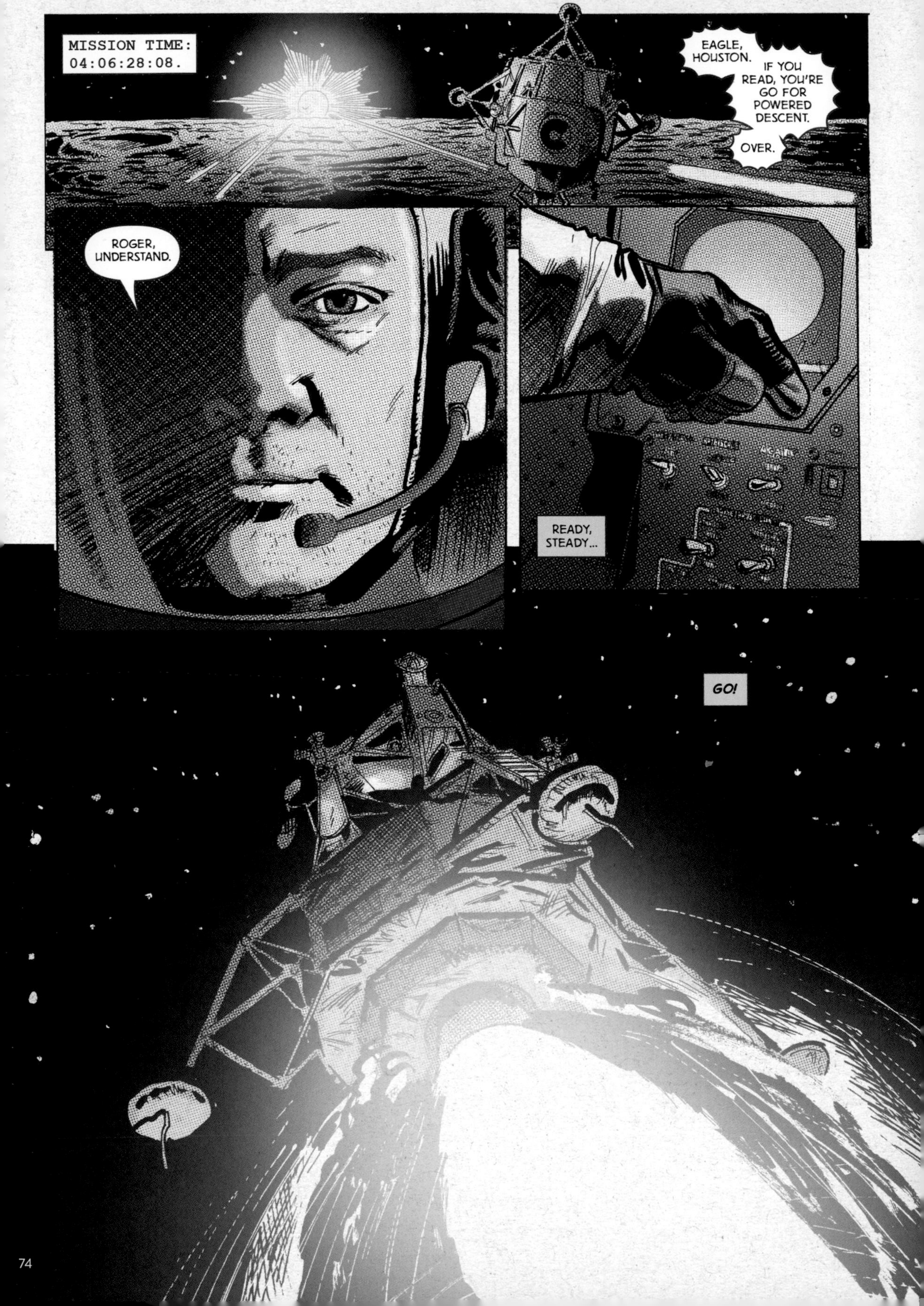
MISSION TIME: 04:06:28:08.
EAGLE, HOUSTON.
IF YOU READ, YOU'RE GO FOR POWERED DESCENT.
OVER.
ROGER, UNDERSTAND.
READY, STEADY...
GO!

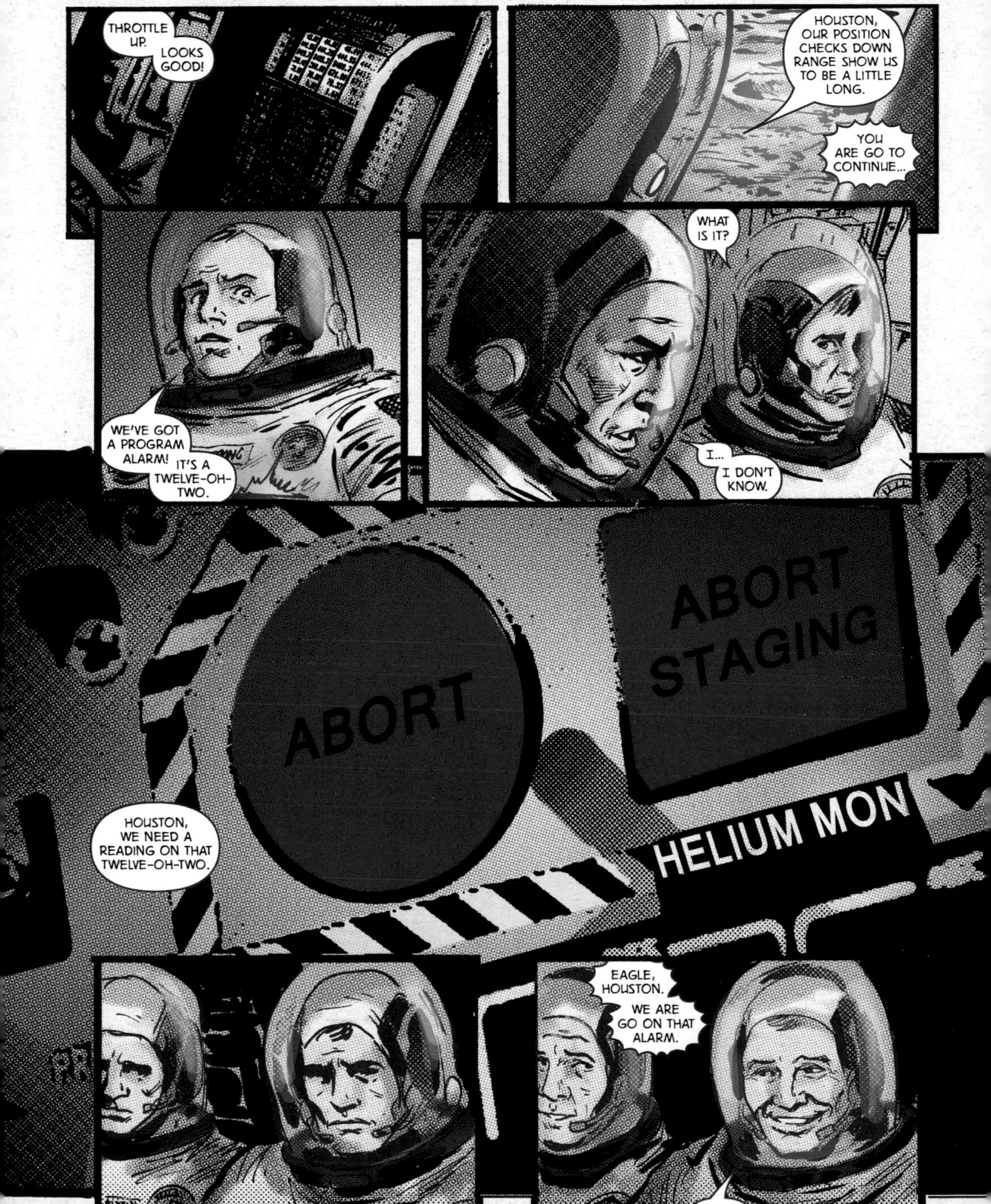
THROTTLE UP.
LOOKS GOOD!
HOUSTON, OUR POSITION CHECKS DOWN RANGE SHOW US TO BE A LITTLE LONG.
YOU ARE GO TO CONTINUE...
WE'VE GOT A PROGRAM ALARM!
IT'S A TWELVE-OH-TWO.
WHAT IS IT?
I... I DON'T KNOW.
ABORT
ABORT STAGING
HELIUM MON
HOUSTON, WE NEED A READING ON THAT TWELVE-OH-TWO.
EAGLE, HOUSTON.
WE ARE GO ON THAT ALARM.
ROGER.

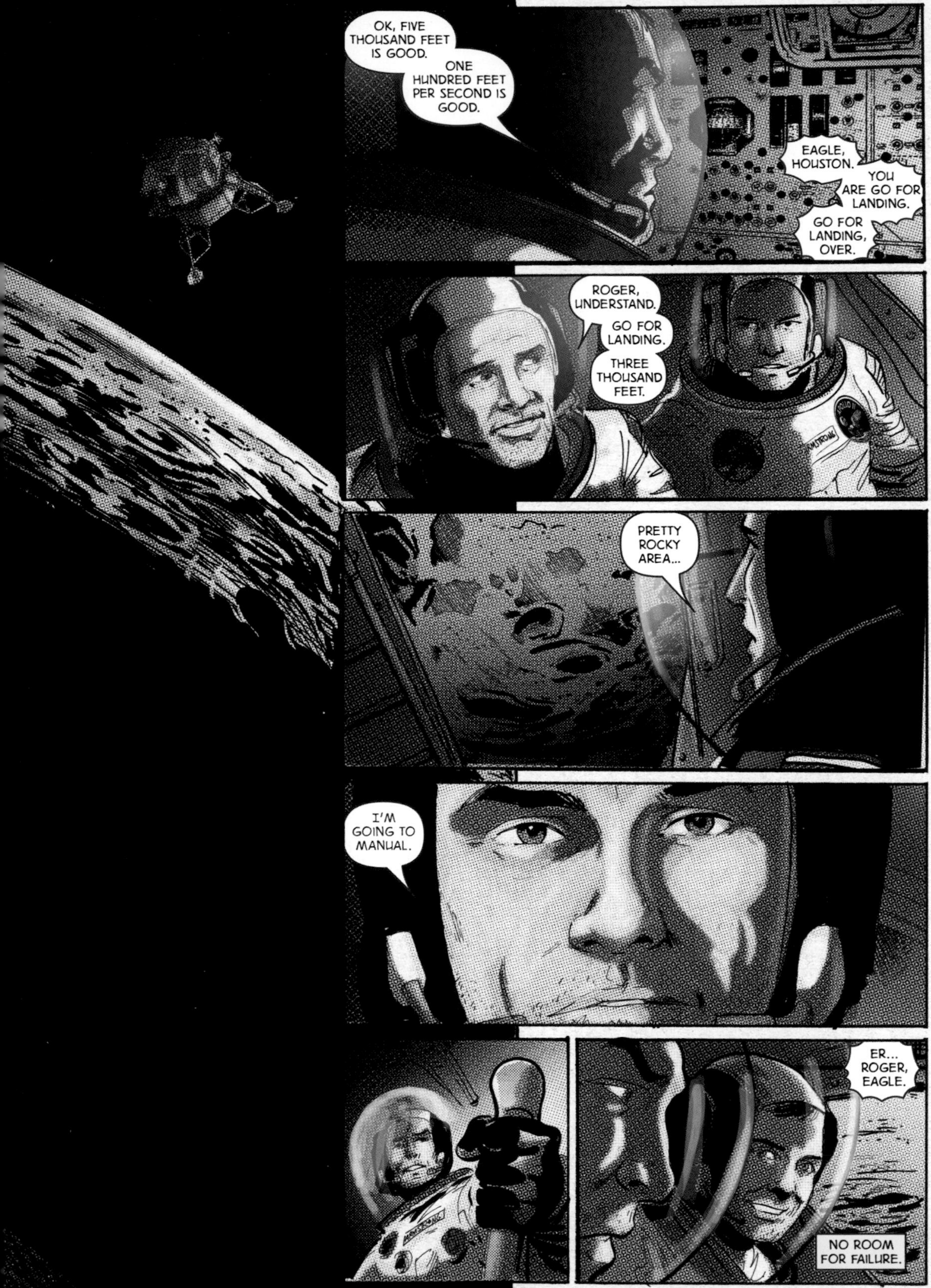
OK, FIVE THOUSAND FEET IS GOOD.
ONE HUNDRED FEET PER SECOND IS GOOD.
EAGLE, HOUSTON.
YOU ARE GO FOR LANDING.
GO FOR LANDING, OVER.
ROGER, UNDERSTAND.
GO FOR LANDING.
THREE THOUSAND FEET.
PRETTY ROCKY AREA...
I'M GOING TO MANUAL.
ER... ROGER, EAGLE.
NO ROOM FOR FAILURE.

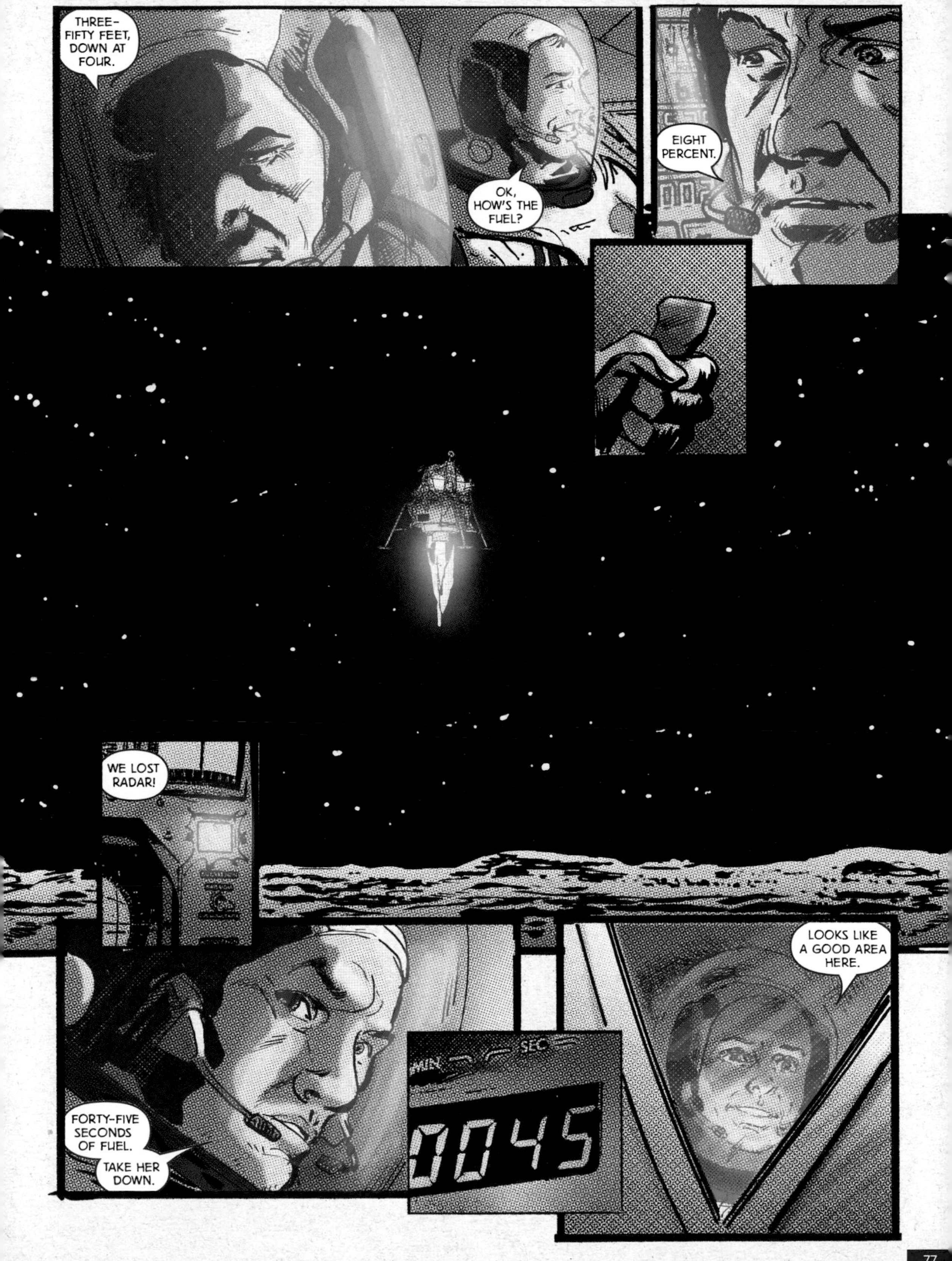
THREE-FIFTY FEET, DOWN AT FOUR.
OK, HOW'S THE FUEL?
EIGHT PERCENT.
WE LOST RADAR!
FORTY-FIVE SECONDS OF FUEL.
TAKE HER DOWN.
LOOKS LIKE A GOOD AREA HERE.

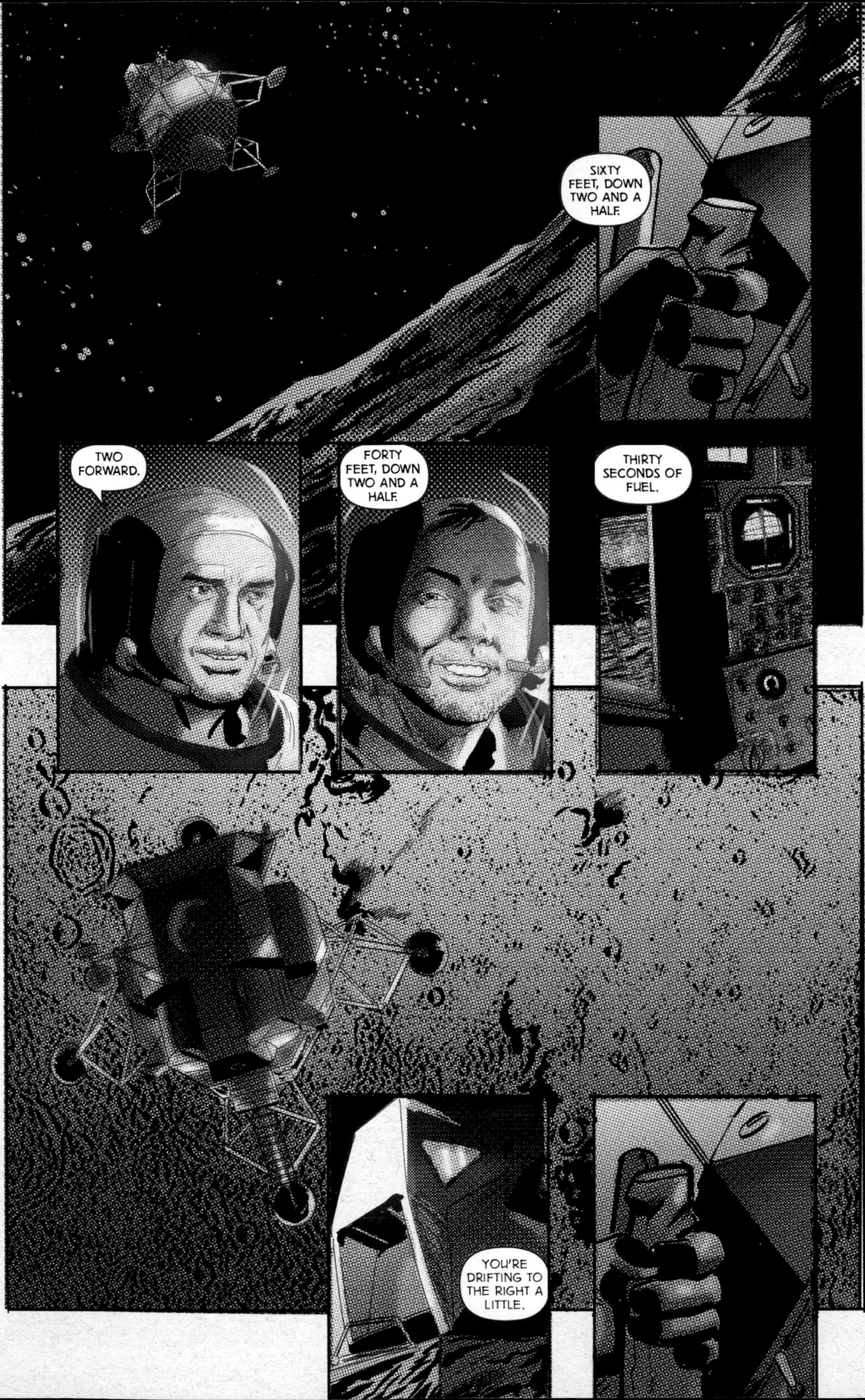
SIXTY FEET, DOWN TWO AND A HALF.
TWO FORWARD.
FORTY FEET, DOWN TWO AND A HALF.
THIRTY SECONDS OF FUEL.
YOU'RE DRIFTING TO THE RIGHT A LITTLE.

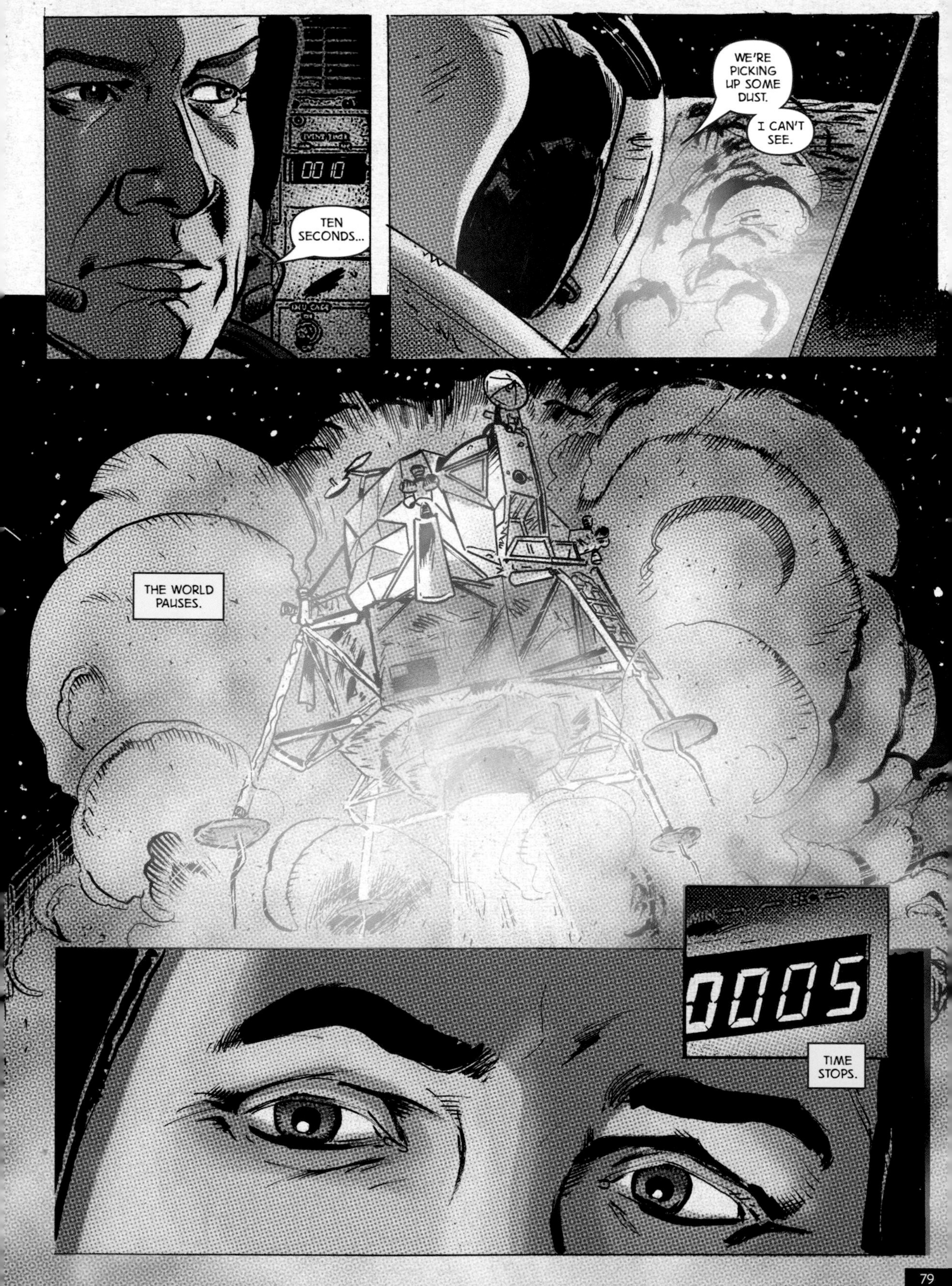
0010
TEN SECONDS...
WE'RE PICKING UP SOME DUST.
I CAN'T SEE.
THE WORLD PAUSES.
0005
TIME STOPS.

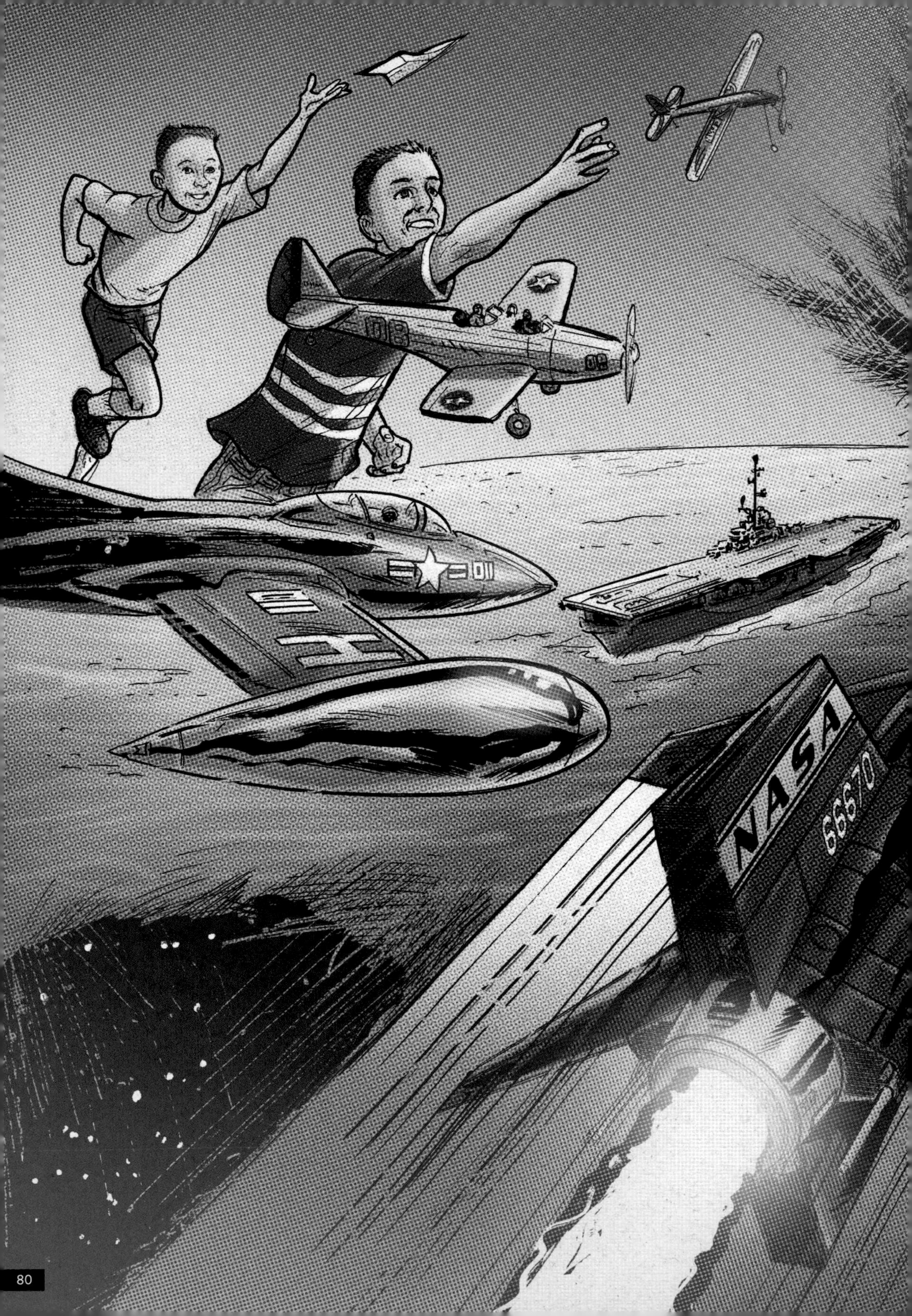

NASA
66670

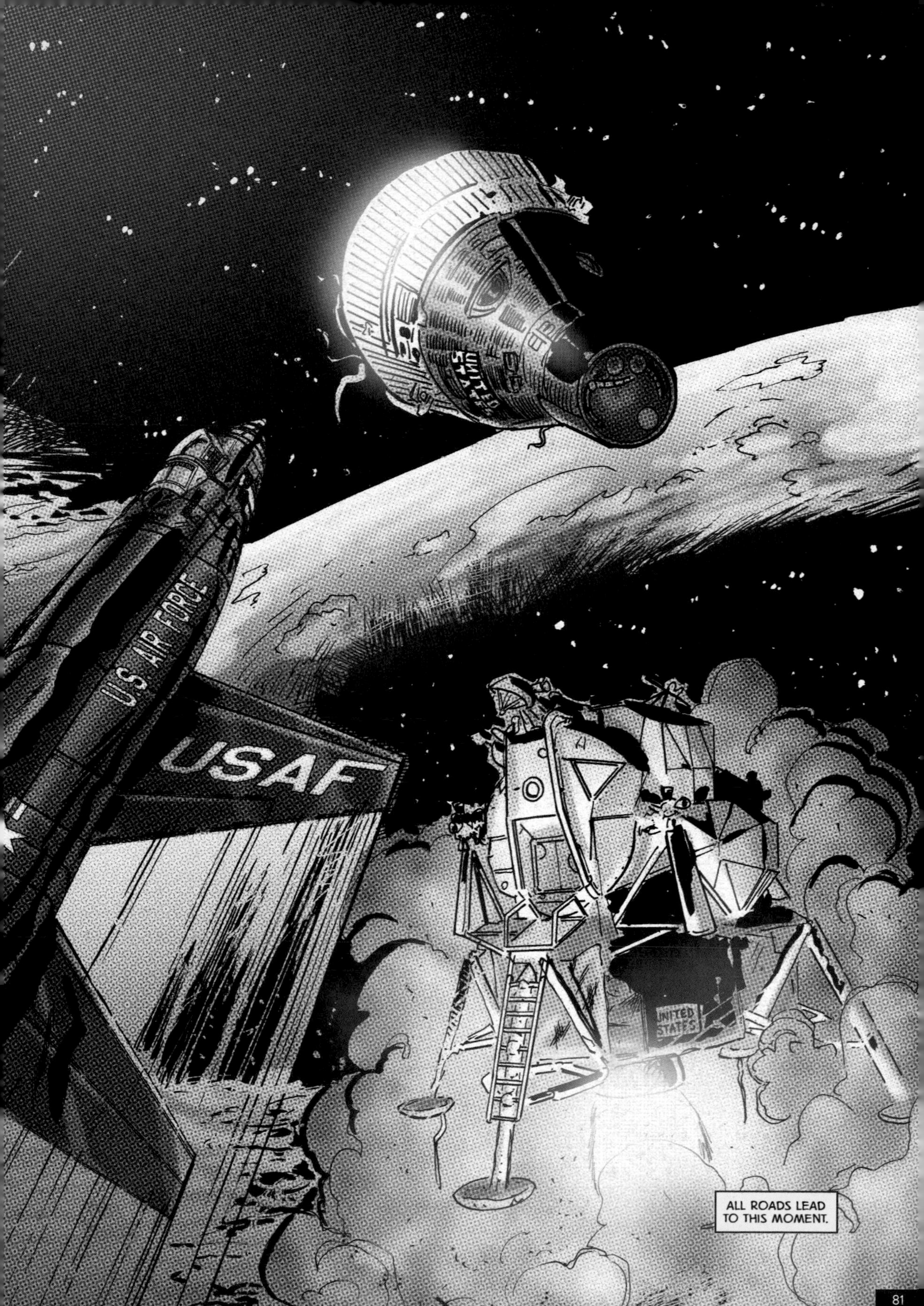
US AIR FORCE
USAF
UNITED STATES
ALL ROADS LEAD TO THIS MOMENT.

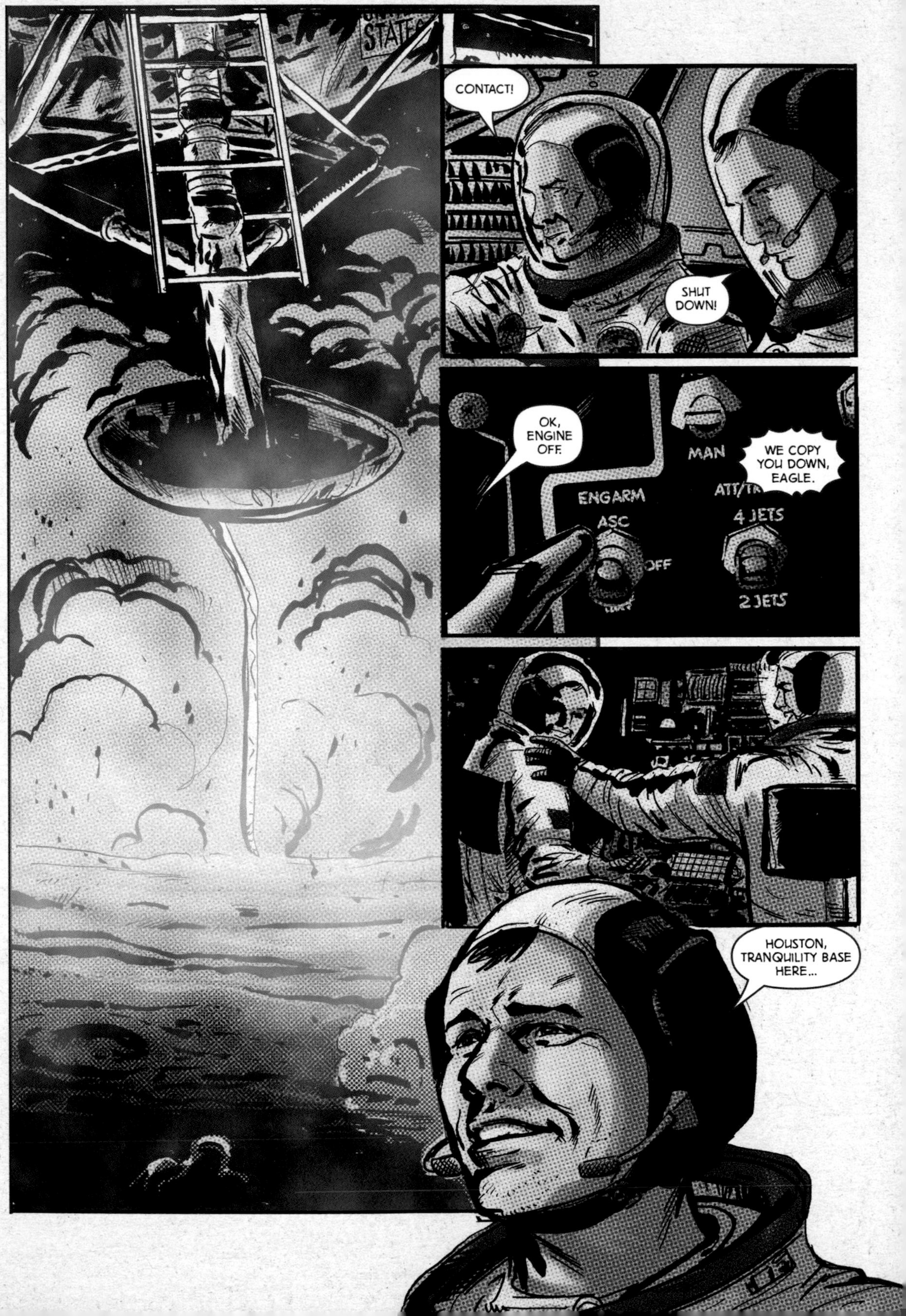
CONTACT!
SHUT DOWN!
OK, ENGINE OFF.
ENGARM
ASC
OFF
MAN
ATT/TR
4 JETS
2 JETS
WE COPY YOU DOWN, EAGLE.
HOUSTON, TRANQUILITY BASE HERE...

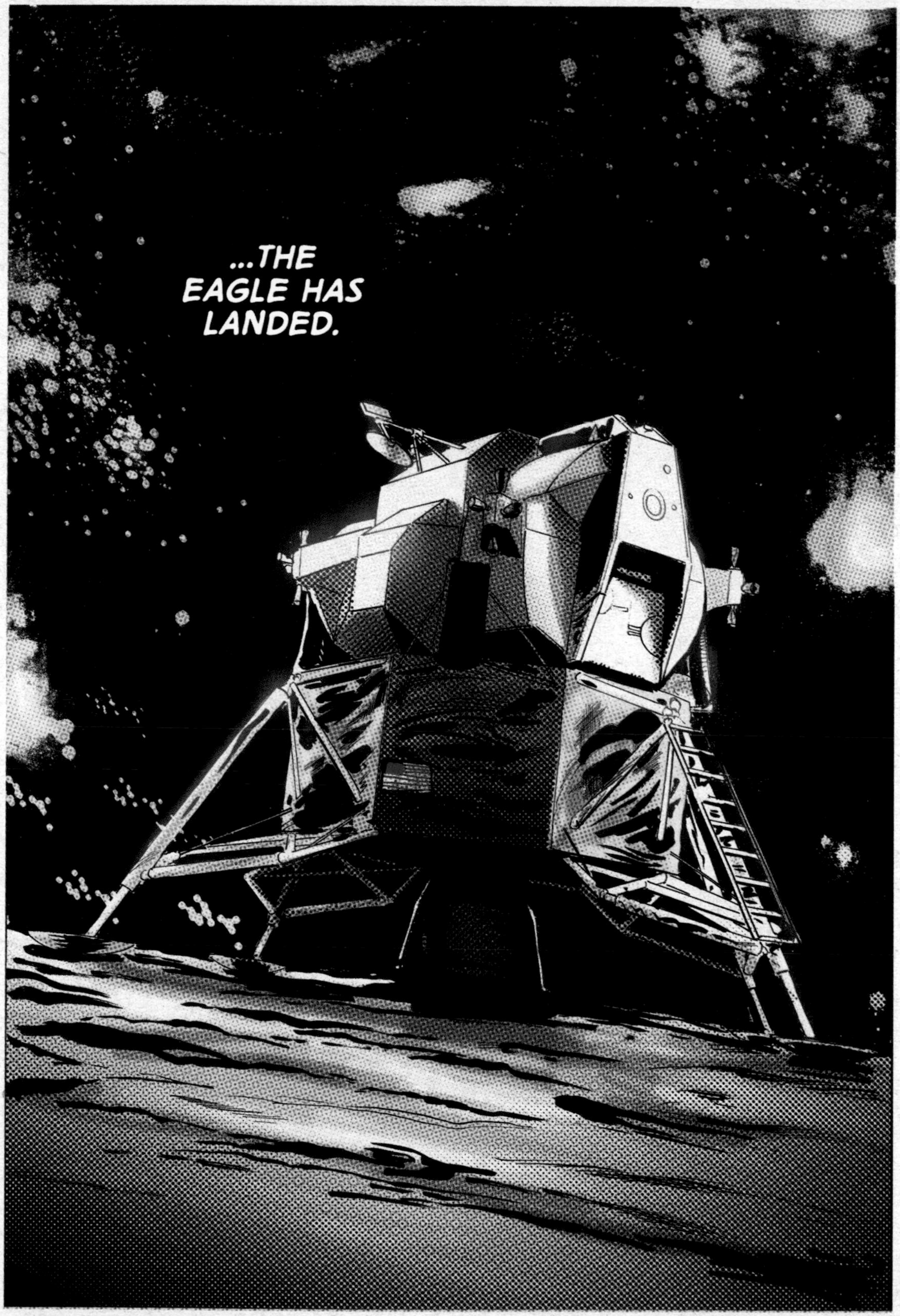
...THE EAGLE HAS LANDED.

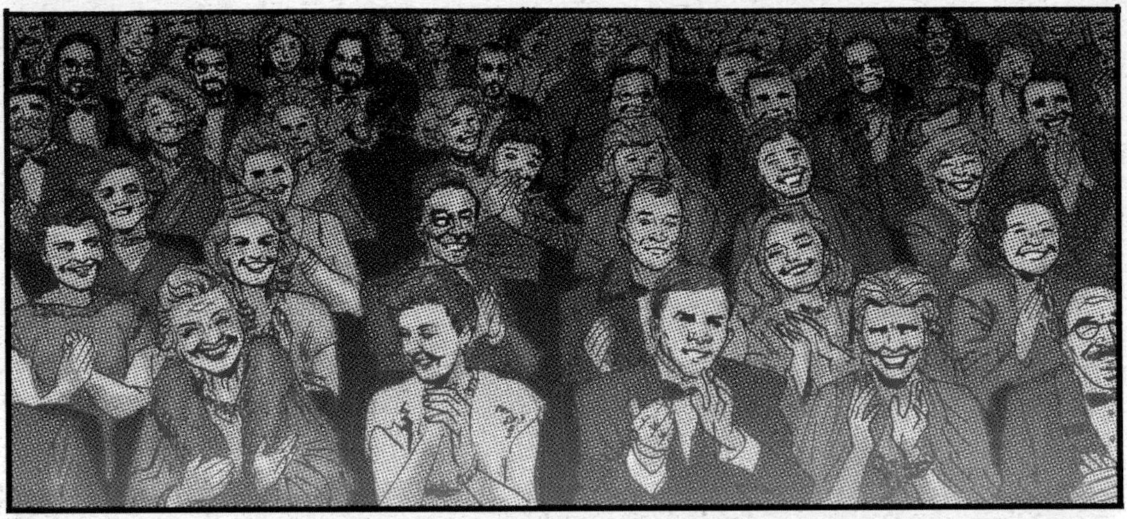

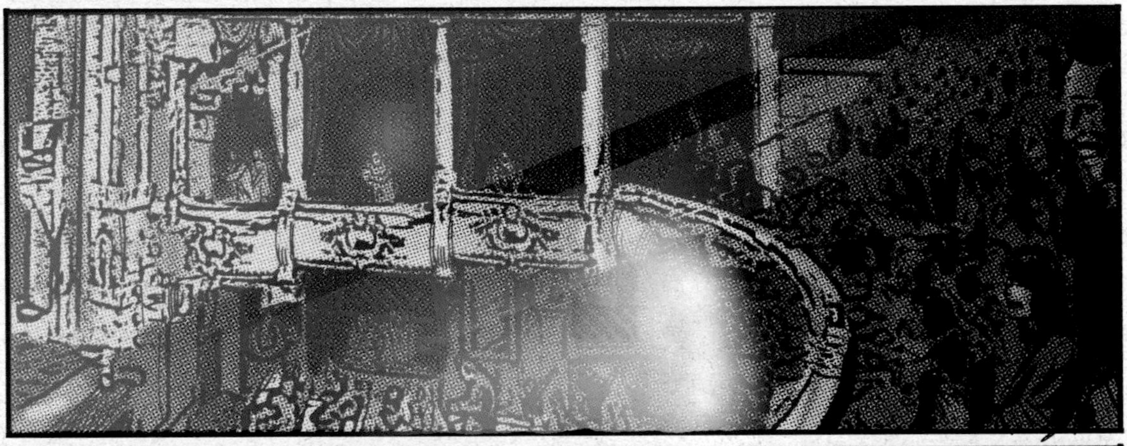

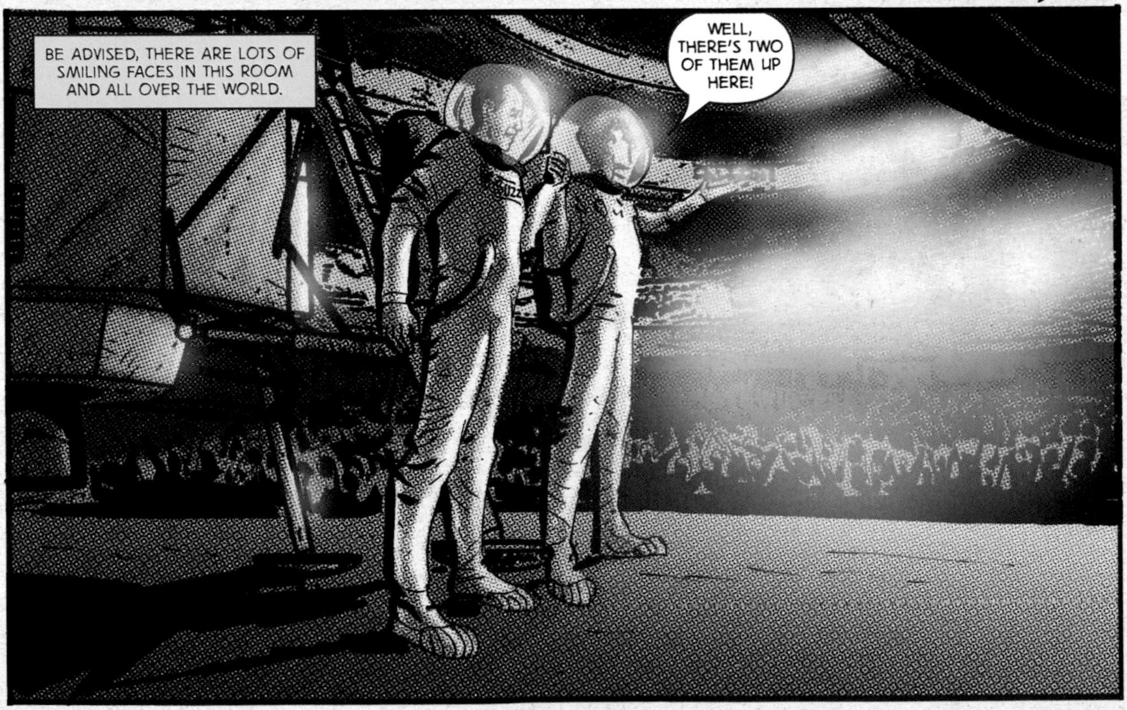
BE ADVISED, THERE ARE LOTS OF SMILING FACES IN THIS ROOM AND ALL OVER THE WORLD.
WELL, THERE'S TWO OF THEM UP HERE!

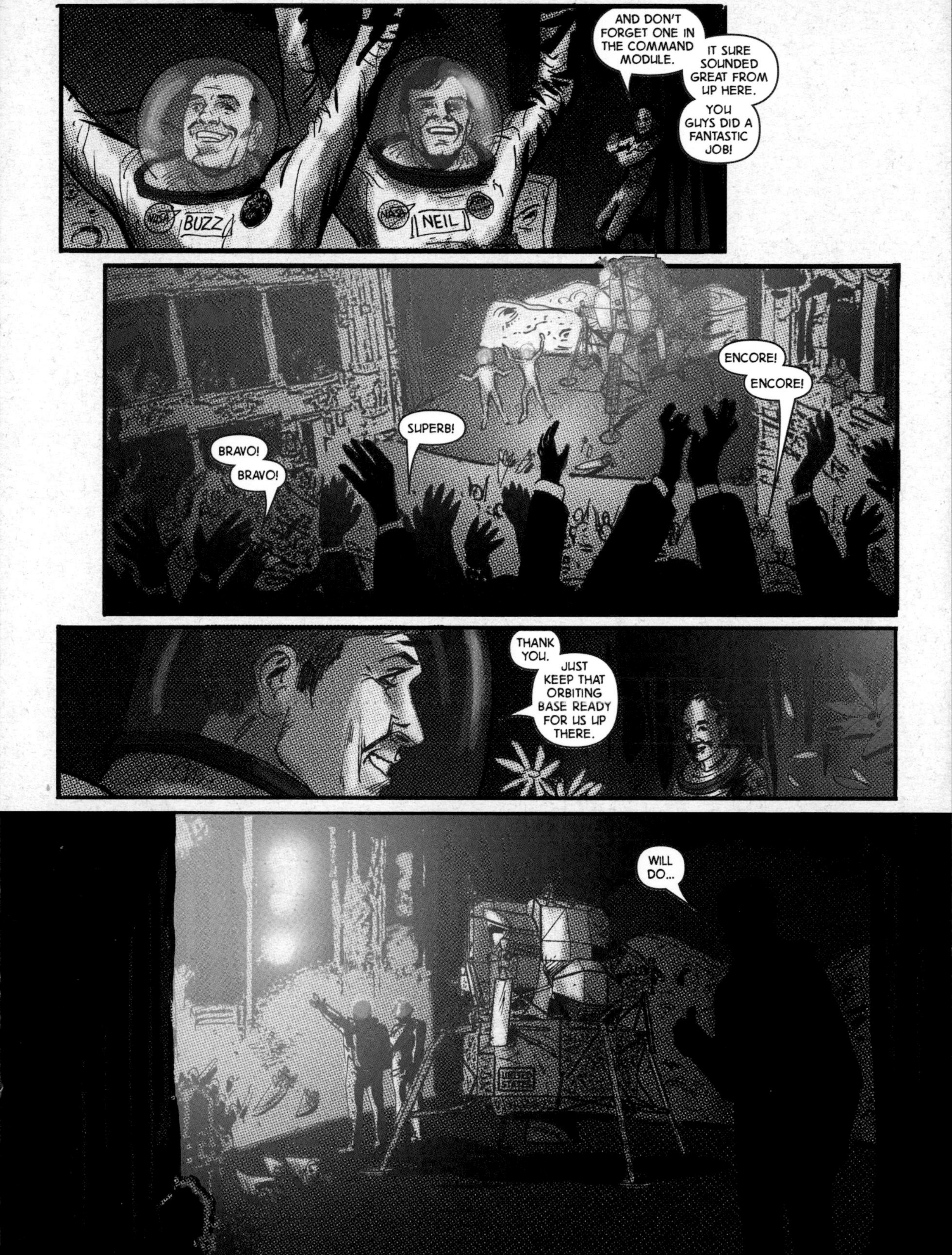
AND DON'T FORGET ONE IN THE COMMAND MODULE.
IT SURE SOUNDED GREAT FROM UP HERE.
YOU GUYS DID A FANTASTIC JOB!
BUZZ
NEIL
BRAVO!
BRAVO!
SUPERB!
ENCORE!
ENCORE!
THANK YOU.
JUST KEEP THAT ORBITING BASE READY FOR US UP THERE.
WILL DO...

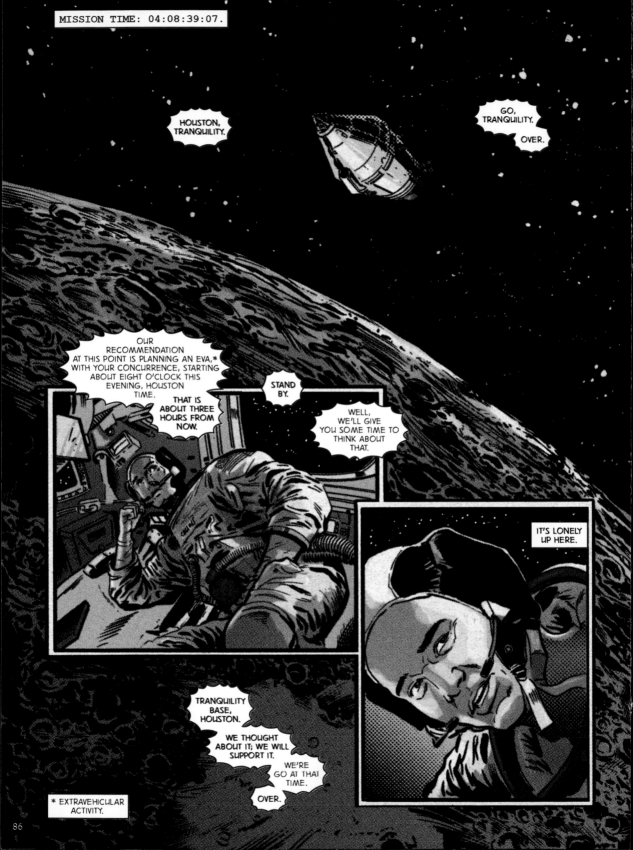

HOUSTON, TRANQUILITY.

GO, TRANQUILITY. OVER.

OUR RECOMMENDATION AT THIS POINT IS PLANNING AN EVA,* WITH YOUR CONCURRENCE, STARTING ABOUT EIGHT O'CLOCK THIS EVENING, HOUSTON TIME.

THAT IS ABOUT THREE HOURS FROM NOW.

STAND BY.

WELL, WE'LL GIVE YOU SOME TIME TO THINK ABOUT THAT.

IT'S LONELY UP HERE.

TRANQUILITY BASE, HOUSTON.

WE THOUGHT ABOUT IT; WE WILL SUPPORT IT.

WE'RE GO AT THAT TIME.

OVER.

* EXTRAVEHICULAR ACTIVITY.

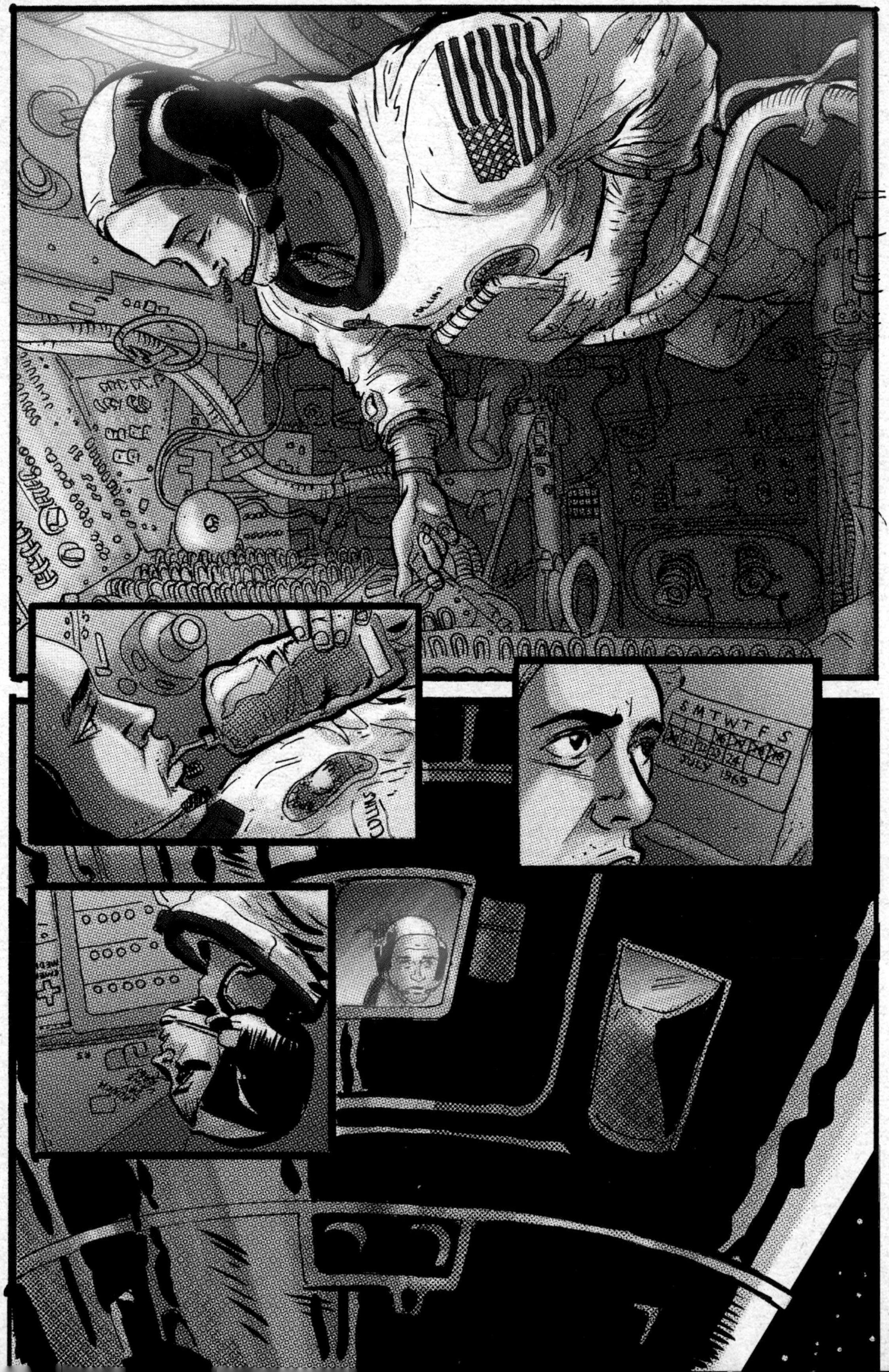
COLLINS
S M T W T F S
JULY 1969

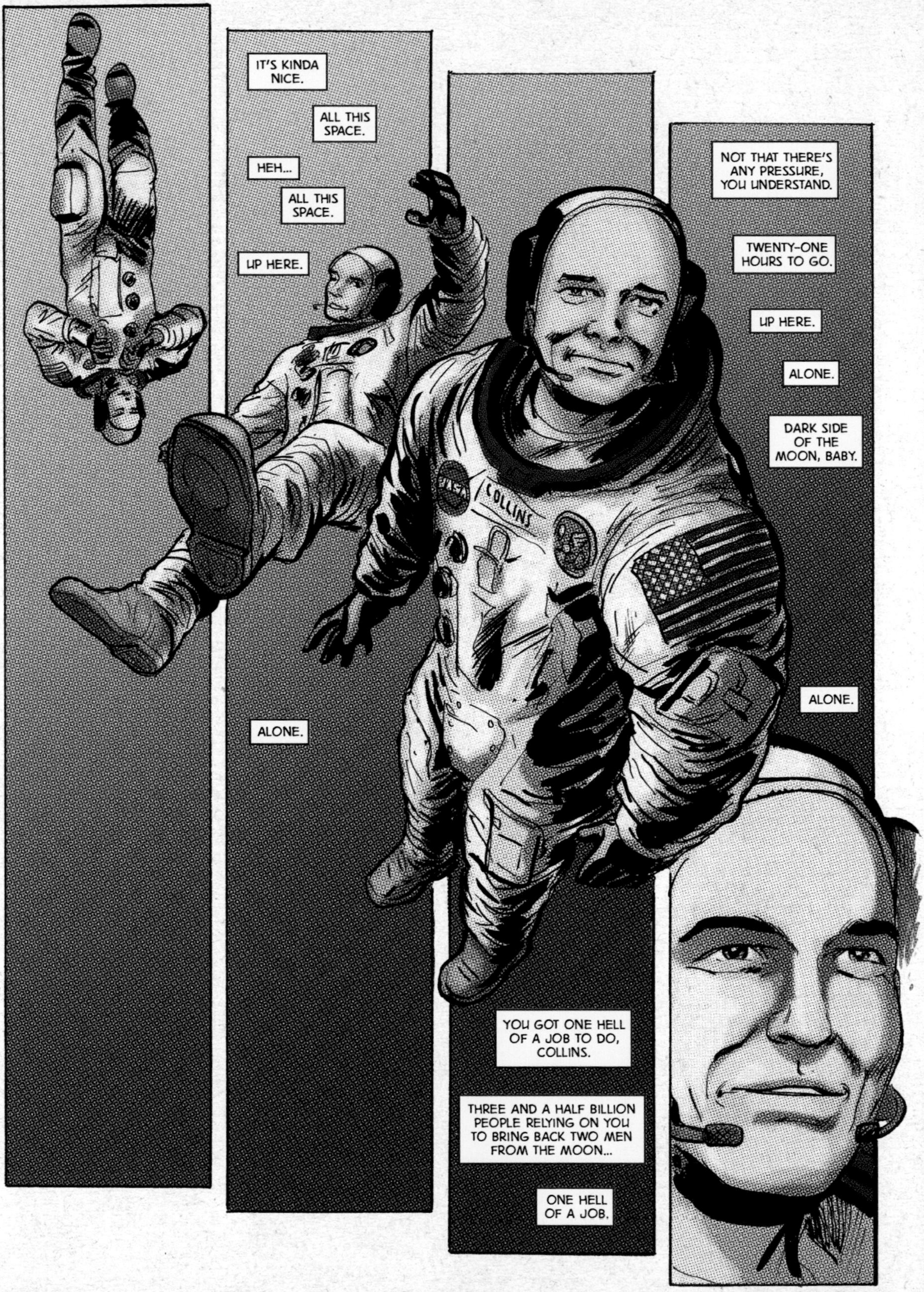
IT'S KINDA NICE.
ALL THIS SPACE.
HEH...
ALL THIS SPACE.
UP HERE.
ALONE.
COLLINS
YOU GOT ONE HELL OF A JOB TO DO, COLLINS.
THREE AND A HALF BILLION PEOPLE RELYING ON YOU TO BRING BACK TWO MEN FROM THE MOON...
ONE HELL OF A JOB.
NOT THAT THERE'S ANY PRESSURE, YOU UNDERSTAND.
TWENTY-ONE HOURS TO GO.
UP HERE.
ALONE.
DARK SIDE OF THE MOON, BABY.
ALONE.

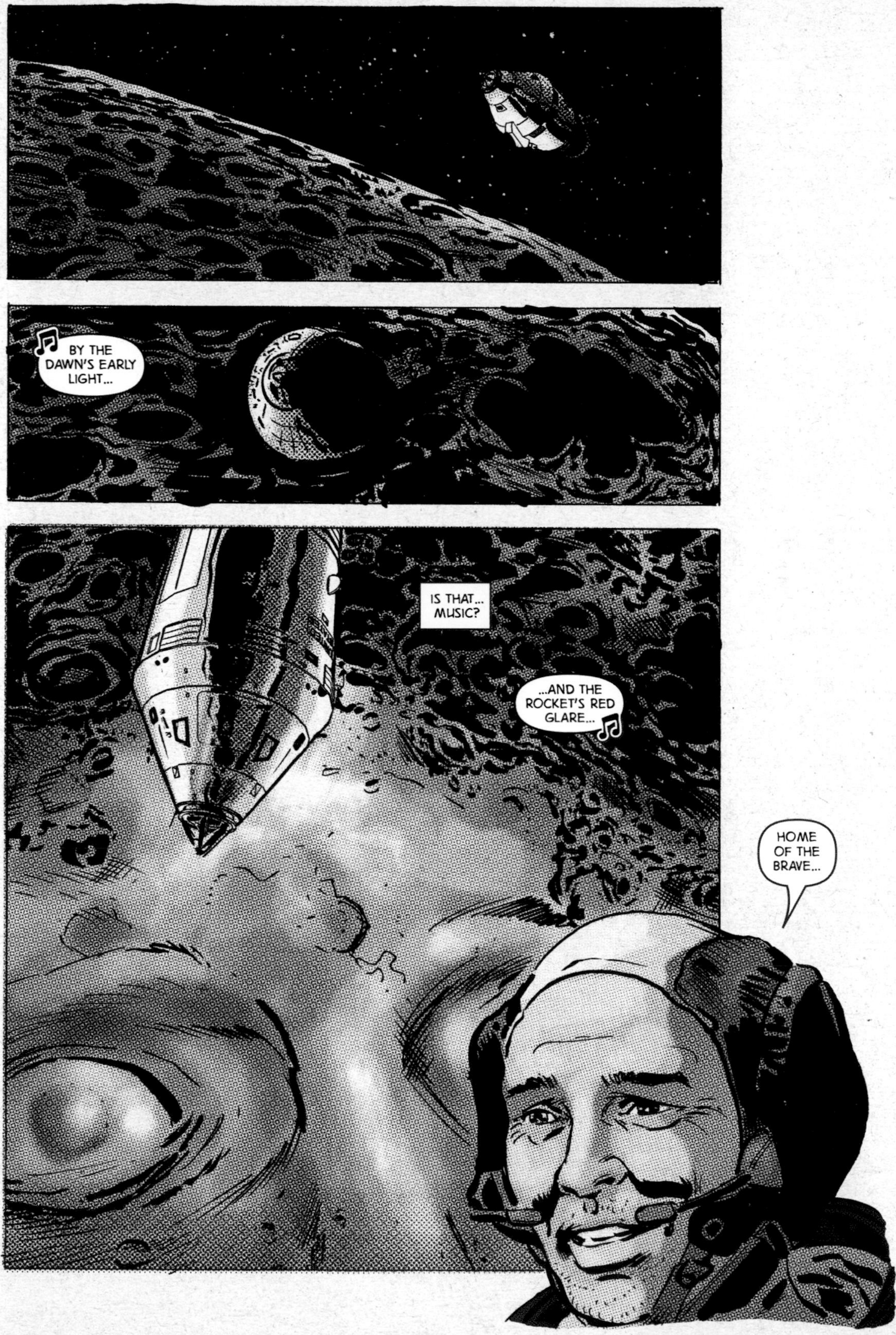

BY THE DAWN'S EARLY LIGHT...
IS THAT... MUSIC?
...AND THE ROCKET'S RED GLARE...
HOME OF THE BRAVE...

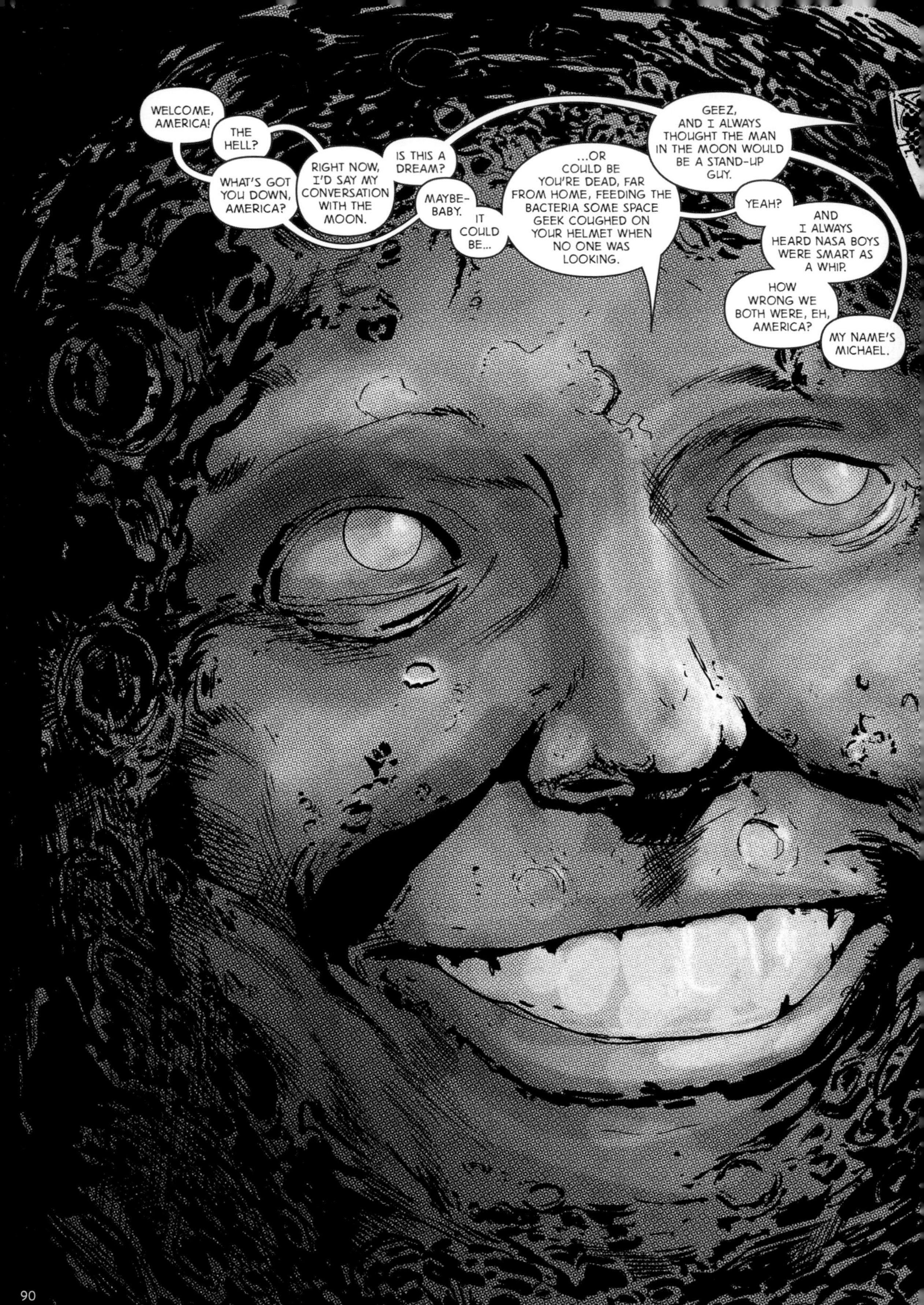
WELCOME, AMERICA!
THE HELL?
WHAT'S GOT YOU DOWN, AMERICA?
RIGHT NOW, I'D SAY MY CONVERSATION WITH THE MOON.
IS THIS A DREAM?
MAYBE-BABY.
IT COULD BE...
...OR COULD BE YOU'RE DEAD, FAR FROM HOME, FEEDING THE BACTERIA SOME SPACE GEEK COUGHED ON YOUR HELMET WHEN NO ONE WAS LOOKING.
GEEZ, AND I ALWAYS THOUGHT THE MAN IN THE MOON WOULD BE A STAND-UP GUY.
YEAH?
AND I ALWAYS HEARD NASA BOYS WERE SMART AS A WHIP.
HOW WRONG WE BOTH WERE, EH, AMERICA?
MY NAME'S MICHAEL.

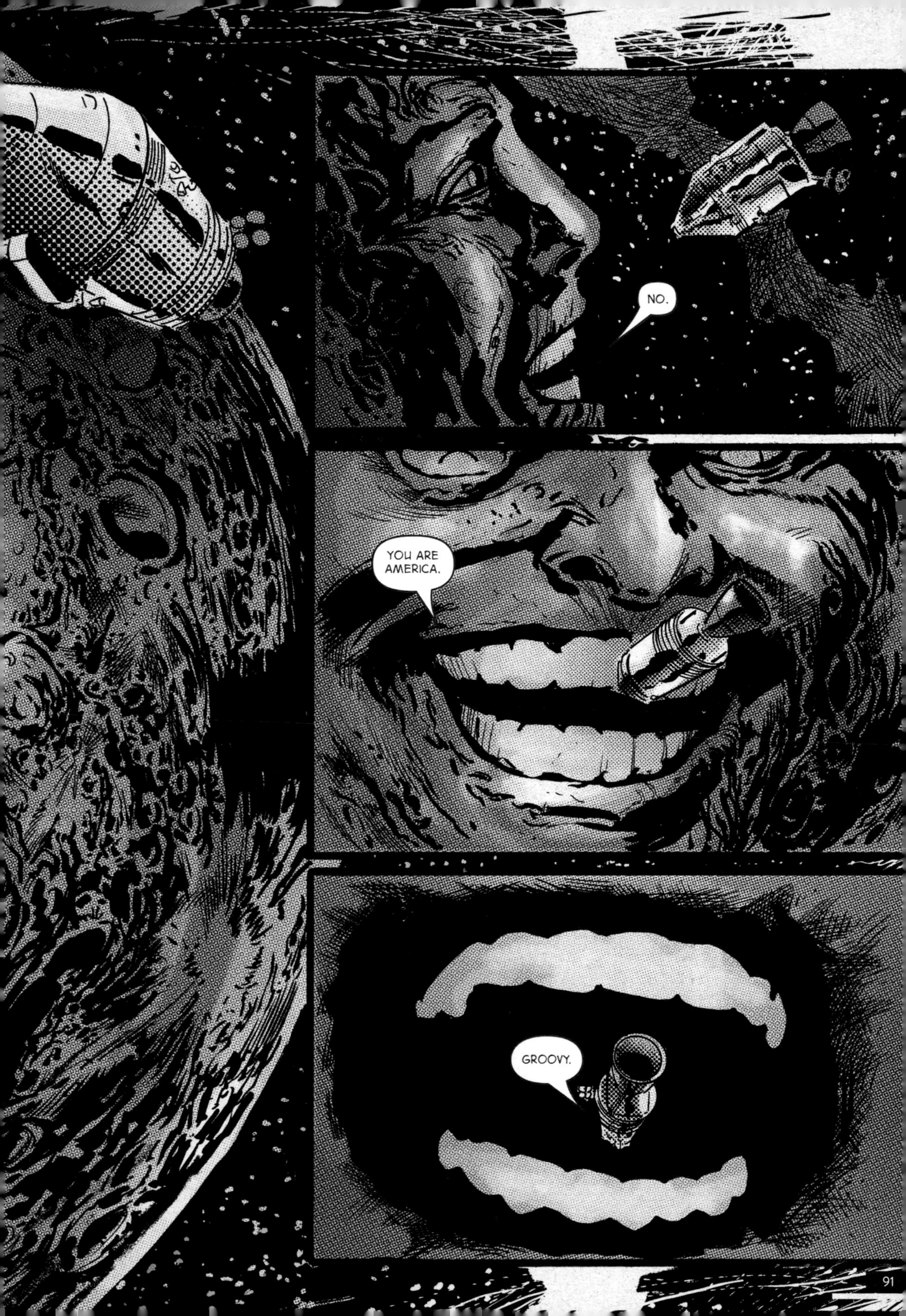
NO.
YOU ARE AMERICA.
GROOVY.

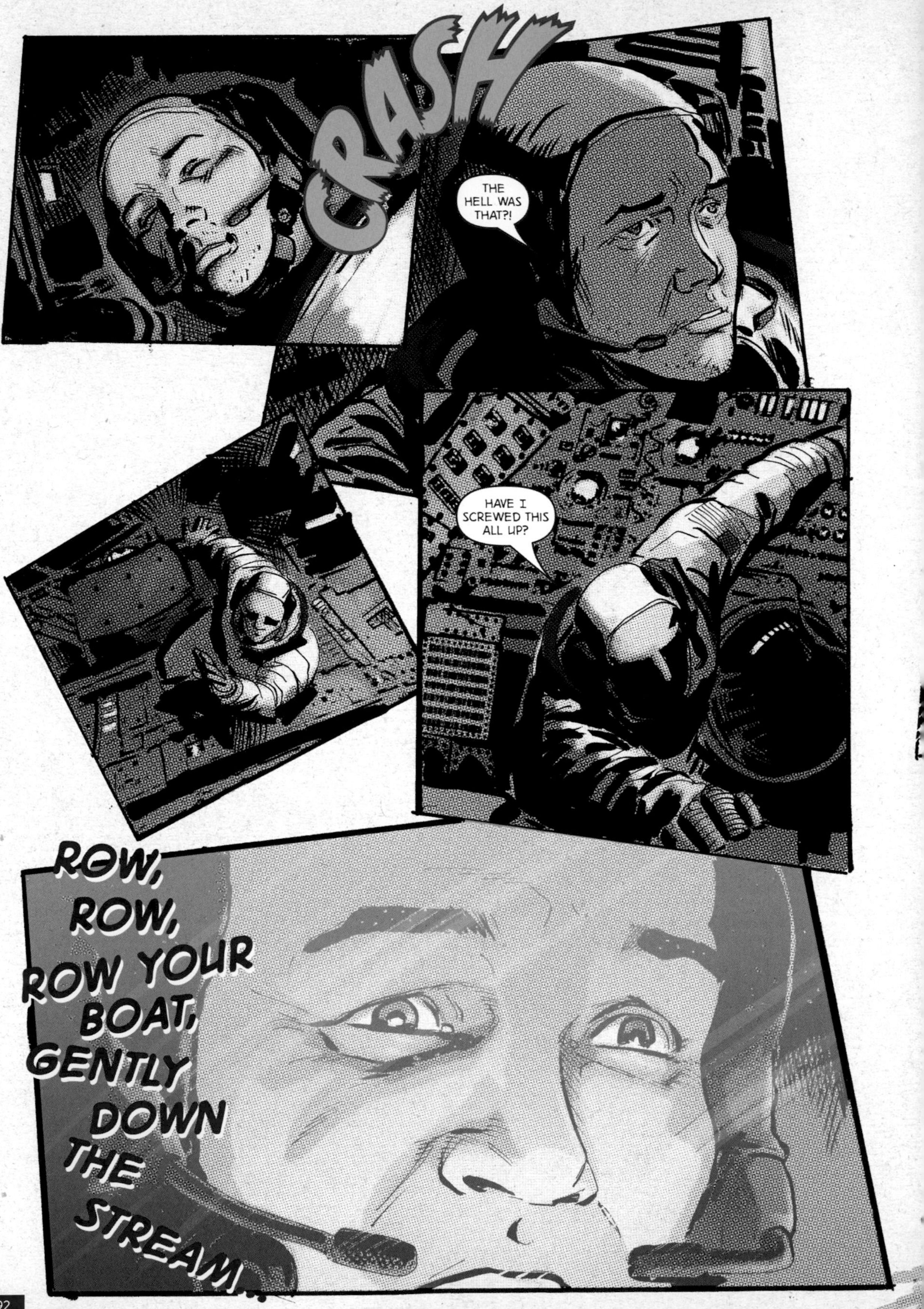
CRASH
THE HELL WAS THAT?!
HAVE I SCREWED THIS ALL UP?
ROW, ROW, ROW YOUR BOAT, GENTLY DOWN THE STREAM...

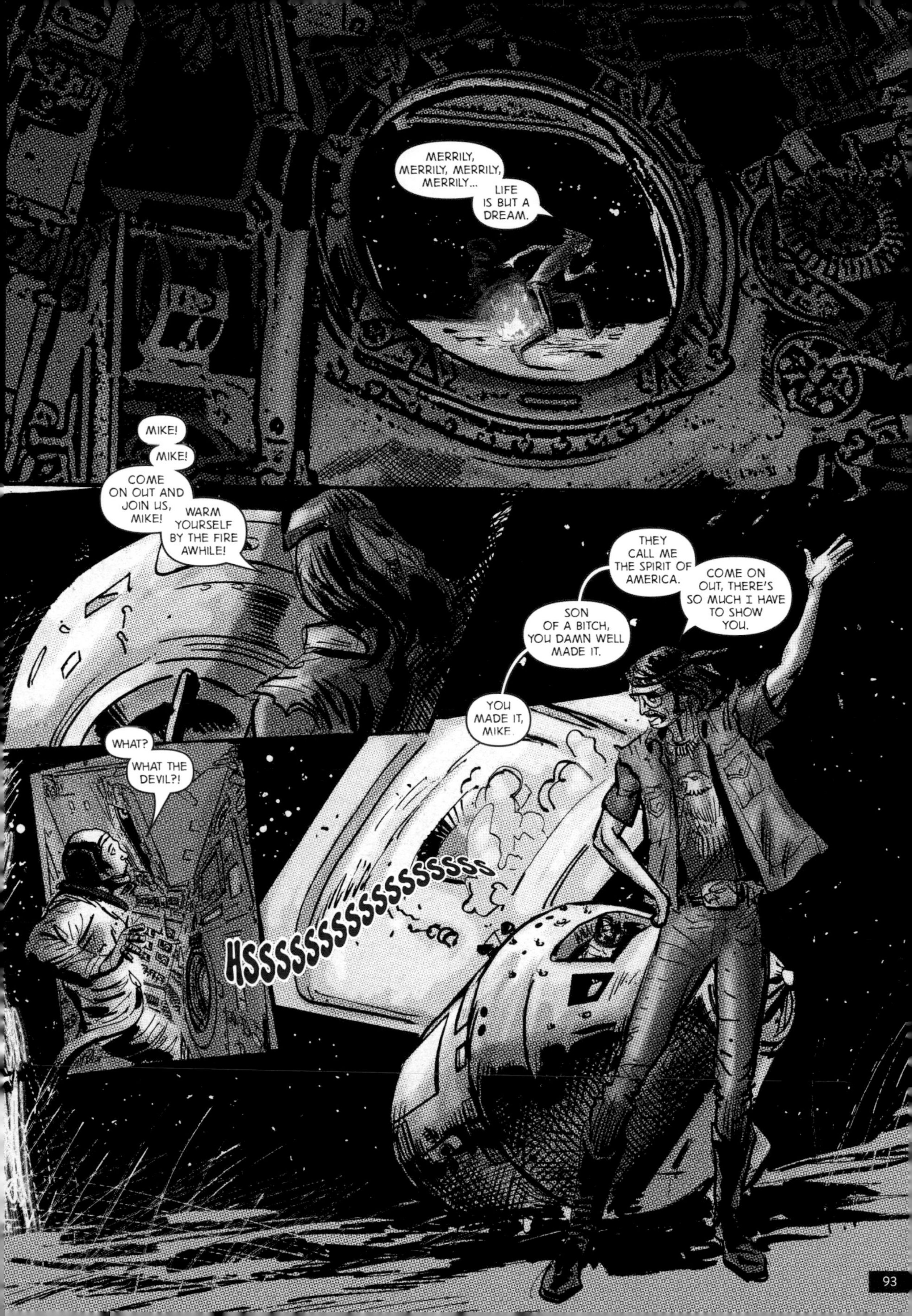
MERRILY, MERRILY, MERRILY, MERRILY...
LIFE IS BUT A DREAM.
MIKE!
MIKE!
COME ON OUT AND JOIN US, MIKE!
WARM YOURSELF BY THE FIRE AWHILE!
THEY CALL ME THE SPIRIT OF AMERICA.
COME ON OUT, THERE'S SO MUCH I HAVE TO SHOW YOU.
SON OF A BITCH, YOU DAMN WELL MADE IT.
YOU MADE IT, MIKE.
WHAT?
WHAT THE DEVIL?!
HSSSSSSSSSSSSSSSSSSS

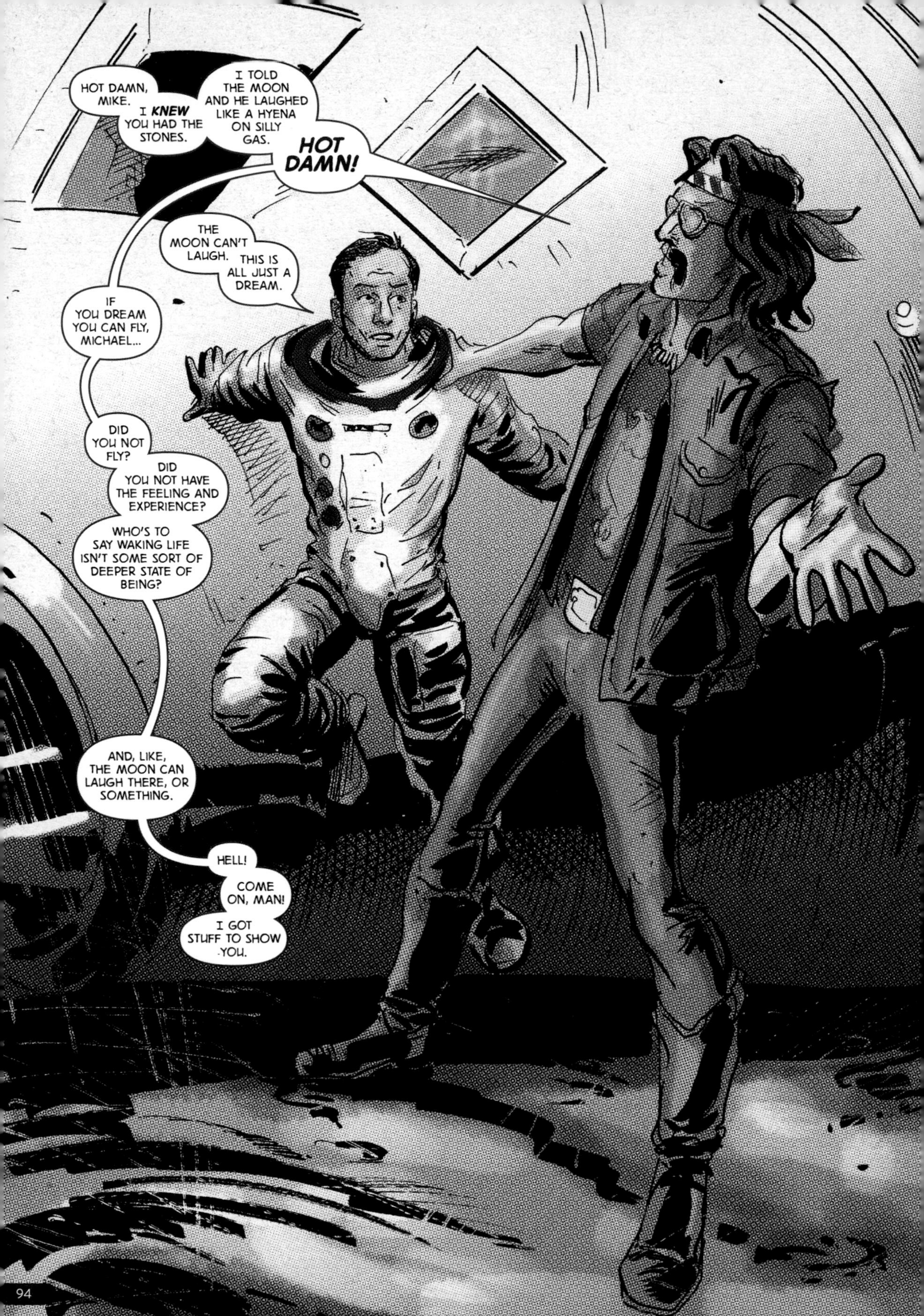
HOT DAMN, MIKE.
I KNEW YOU HAD THE STONES.
I TOLD THE MOON AND HE LAUGHED LIKE A HYENA ON SILLY GAS.
HOT DAMN!
THE MOON CAN'T LAUGH.
THIS IS ALL JUST A DREAM.
IF YOU DREAM YOU CAN FLY, MICHAEL...
DID YOU NOT FLY?
DID YOU NOT HAVE THE FEELING AND EXPERIENCE?
WHO'S TO SAY WAKING LIFE ISN'T SOME SORT OF DEEPER STATE OF BEING?
AND, LIKE, THE MOON CAN LAUGH THERE, OR SOMETHING.
HELL!
COME ON, MAN!
I GOT STUFF TO SHOW YOU.

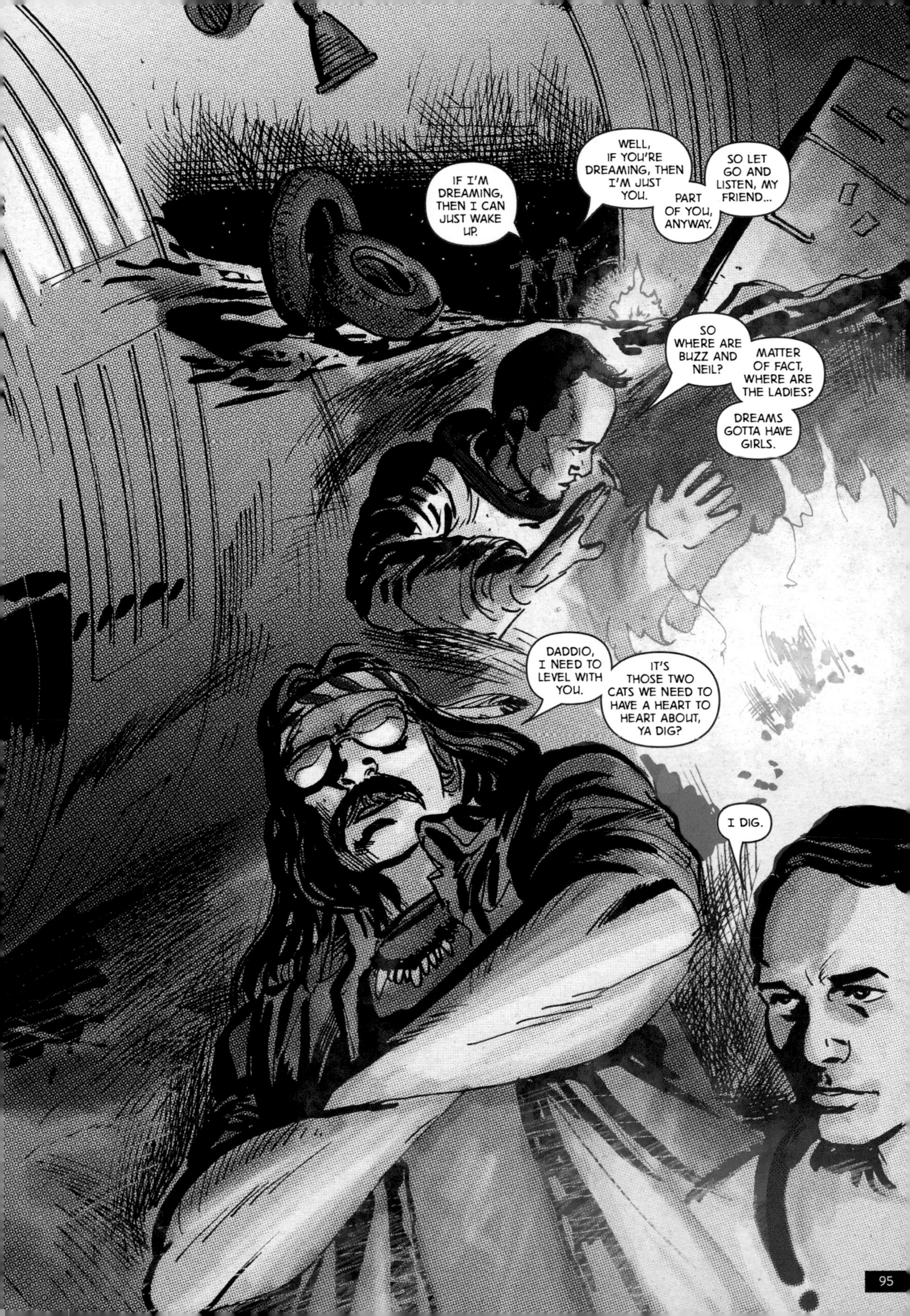
IF I'M DREAMING, THEN I CAN JUST WAKE UP.
WELL, IF YOU'RE DREAMING, THEN I'M JUST YOU.
PART OF YOU, ANYWAY.
SO LET GO AND LISTEN, MY FRIEND...
SO WHERE ARE BUZZ AND NEIL?
MATTER OF FACT, WHERE ARE THE LADIES?
DREAMS GOTTA HAVE GIRLS.
DADDIO, I NEED TO LEVEL WITH YOU.
IT'S THOSE TWO CATS WE NEED TO HAVE A HEART TO HEART ABOUT, YA DIG?
I DIG.

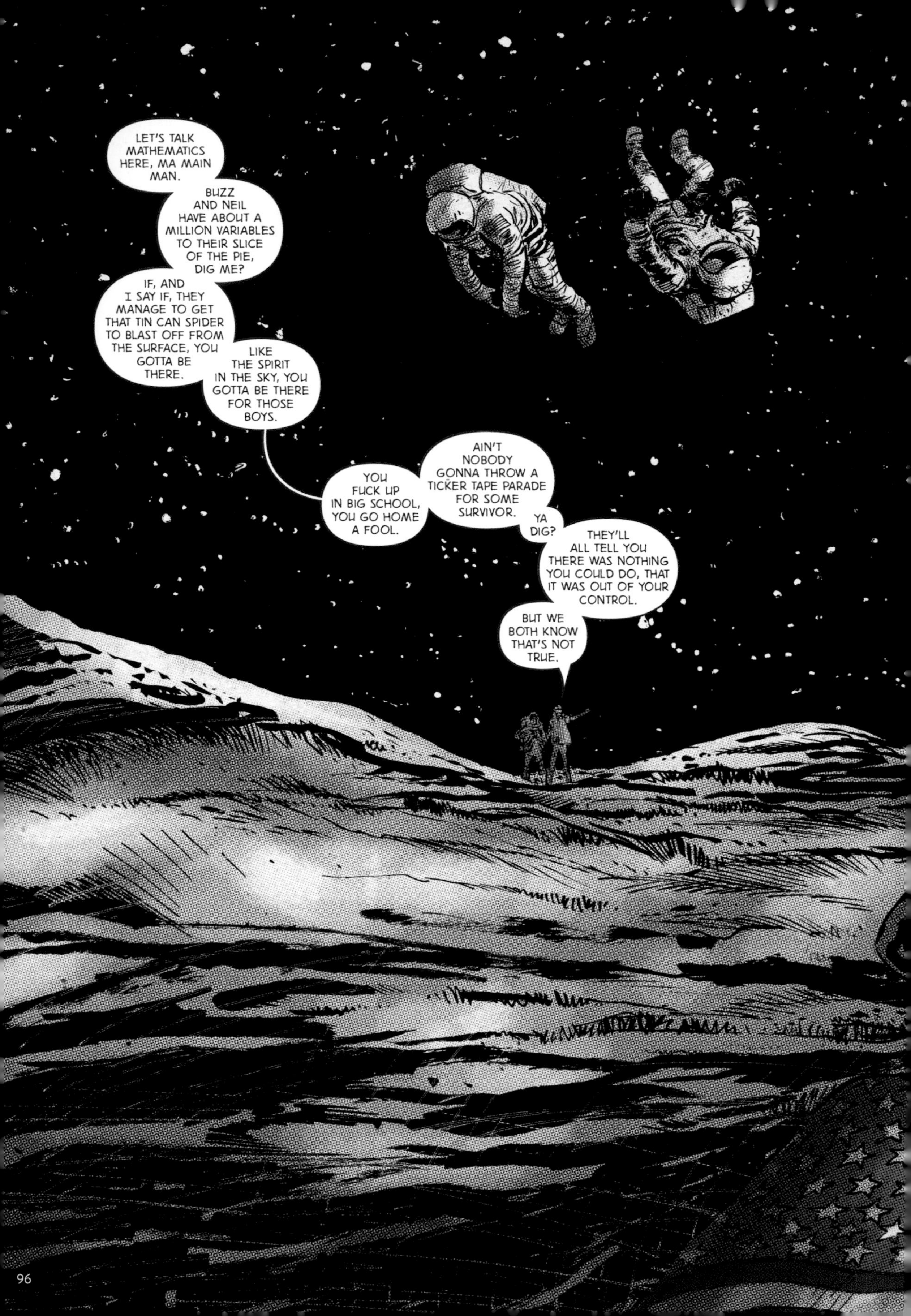
LET'S TALK MATHEMATICS HERE, MA MAIN MAN.
BUZZ AND NEIL HAVE ABOUT A MILLION VARIABLES TO THEIR SLICE OF THE PIE, DIG ME?
IF, AND I SAY IF, THEY MANAGE TO GET THAT TIN CAN SPIDER TO BLAST OFF FROM THE SURFACE, YOU GOTTA BE THERE.
LIKE THE SPIRIT IN THE SKY, YOU GOTTA BE THERE FOR THOSE BOYS.
YOU FUCK UP IN BIG SCHOOL, YOU GO HOME A FOOL.
AIN'T NOBODY GONNA THROW A TICKER TAPE PARADE FOR SOME SURVIVOR.
YA DIG?
THEY'LL ALL TELL YOU THERE WAS NOTHING YOU COULD DO, THAT IT WAS OUT OF YOUR CONTROL.
BUT WE BOTH KNOW THAT'S NOT TRUE.

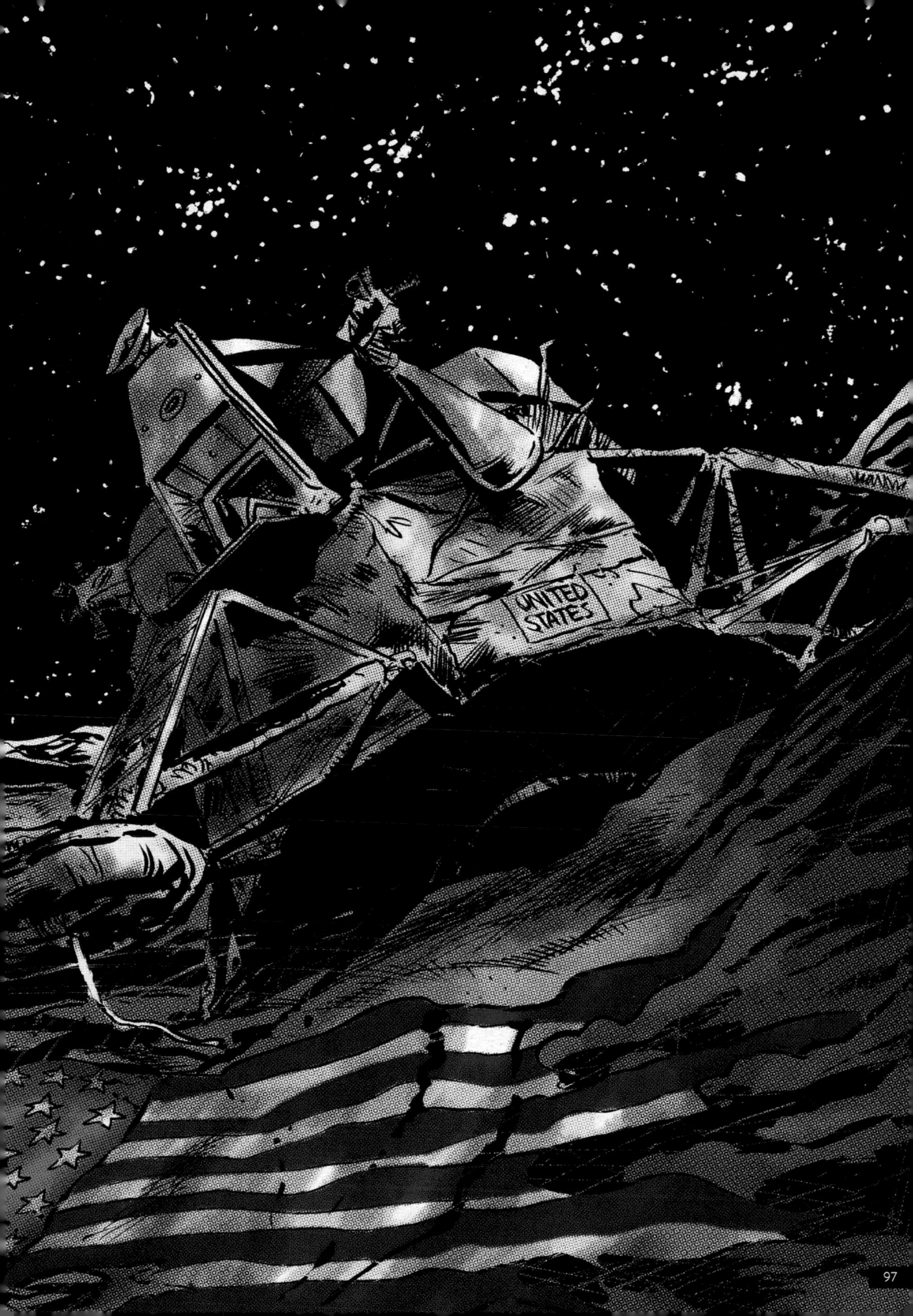
UNITED STATES

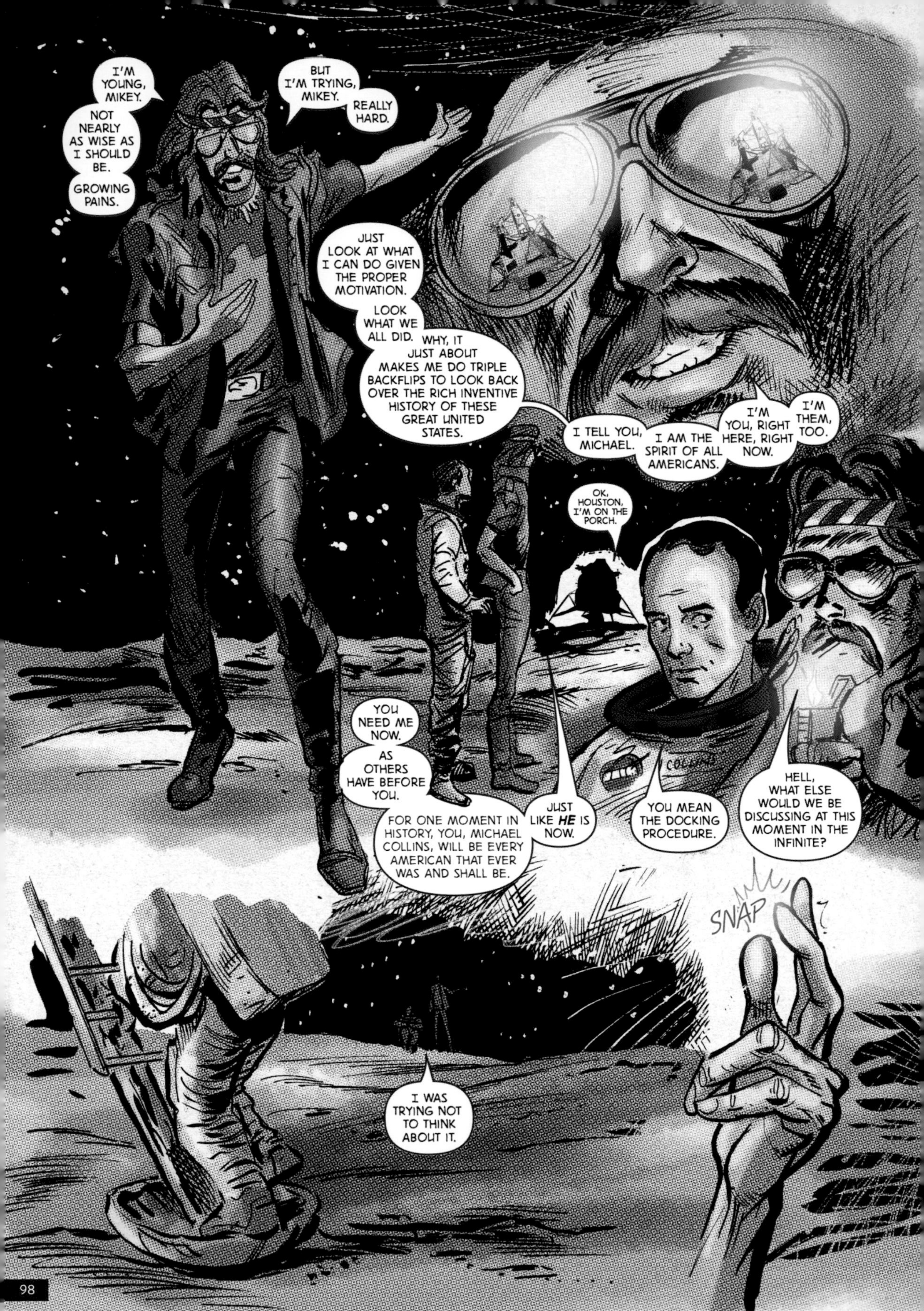
I'M YOUNG, MIKEY.
NOT NEARLY AS WISE AS I SHOULD BE.
GROWING PAINS.
BUT I'M TRYING, MIKEY.
REALLY HARD.
JUST LOOK AT WHAT I CAN DO GIVEN THE PROPER MOTIVATION.
LOOK WHAT WE ALL DID.
WHY, IT JUST ABOUT MAKES ME DO TRIPLE BACKFLIPS TO LOOK BACK OVER THE RICH INVENTIVE HISTORY OF THESE GREAT UNITED STATES.
I TELL YOU, MICHAEL.
I AM THE SPIRIT OF ALL AMERICANS.
I'M YOU, RIGHT HERE, RIGHT NOW.
I'M THEM, TOO.
OK, HOUSTON, I'M ON THE PORCH.
YOU NEED ME NOW.
AS OTHERS HAVE BEFORE YOU.
FOR ONE MOMENT IN HISTORY, YOU, MICHAEL COLLINS, WILL BE EVERY AMERICAN THAT EVER WAS AND SHALL BE.
JUST LIKE HE IS NOW.
YOU MEAN THE DOCKING PROCEDURE.
HELL, WHAT ELSE WOULD WE BE DISCUSSING AT THIS MOMENT IN THE INFINITE?
SNAP
I WAS TRYING NOT TO THINK ABOUT IT.

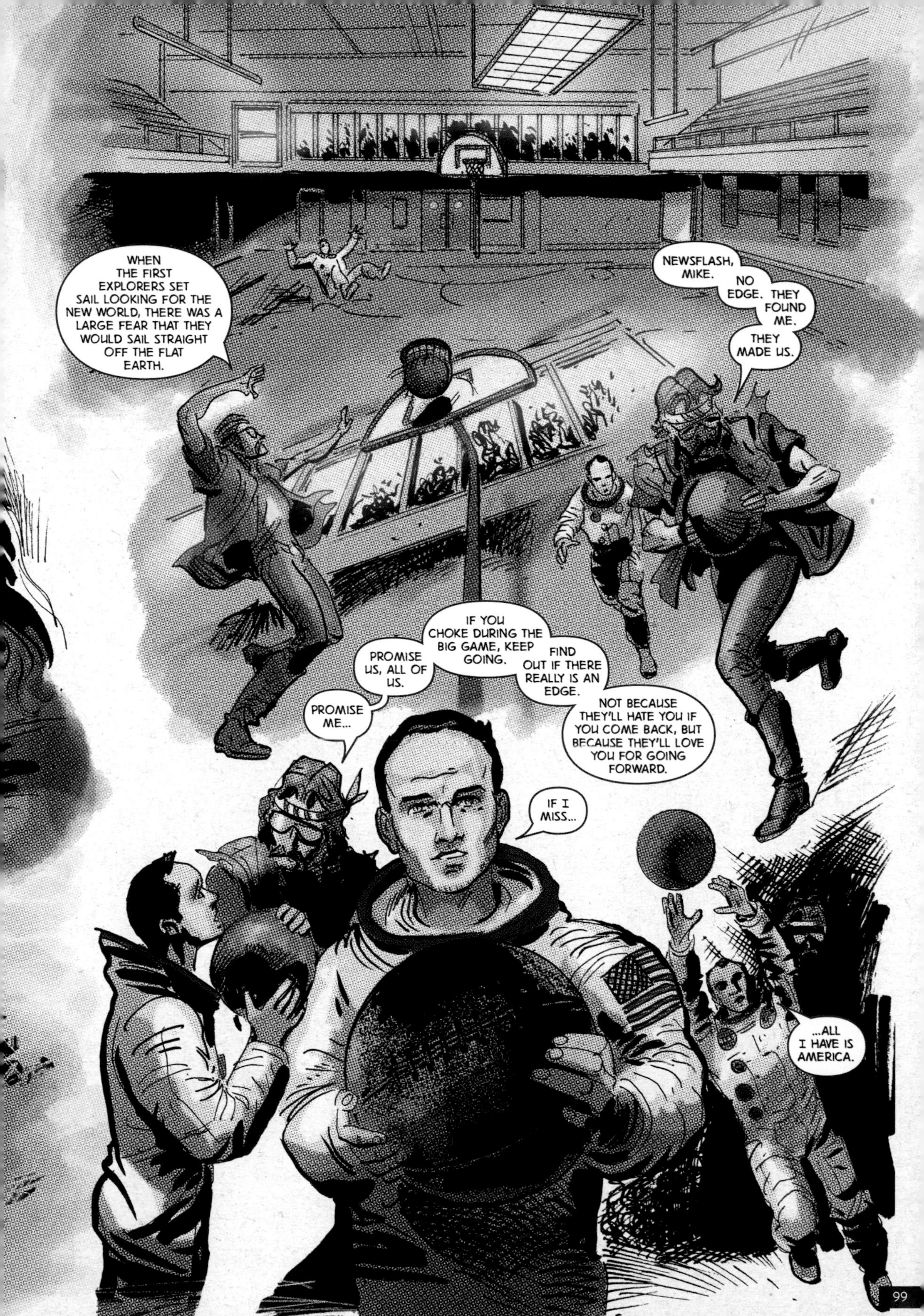
WHEN THE FIRST EXPLORERS SET SAIL LOOKING FOR THE NEW WORLD, THERE WAS A LARGE FEAR THAT THEY WOULD SAIL STRAIGHT OFF THE FLAT EARTH.
NEWSFLASH, MIKE.
NO EDGE. THEY FOUND ME.
THEY MADE US.
PROMISE ME...
PROMISE US, ALL OF US.
IF YOU CHOKE DURING THE BIG GAME, KEEP GOING.
FIND OUT IF THERE REALLY IS AN EDGE.
NOT BECAUSE THEY'LL HATE YOU IF YOU COME BACK, BUT BECAUSE THEY'LL LOVE YOU FOR GOING FORWARD.
IF I MISS...
...ALL I HAVE IS AMERICA.

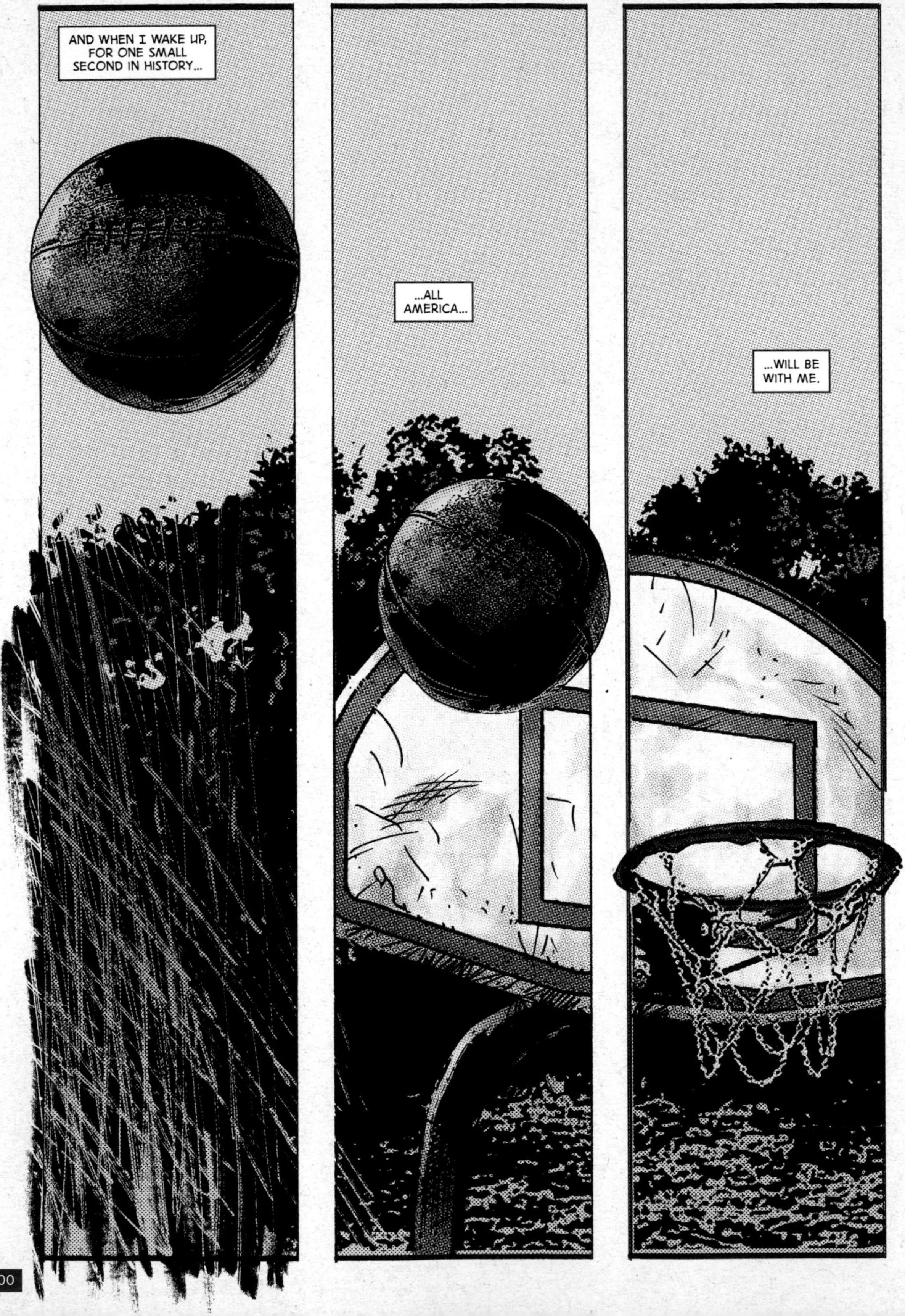
AND WHEN I WAKE UP, FOR ONE SMALL SECOND IN HISTORY...
...ALL AMERICA...
...WILL BE WITH ME.

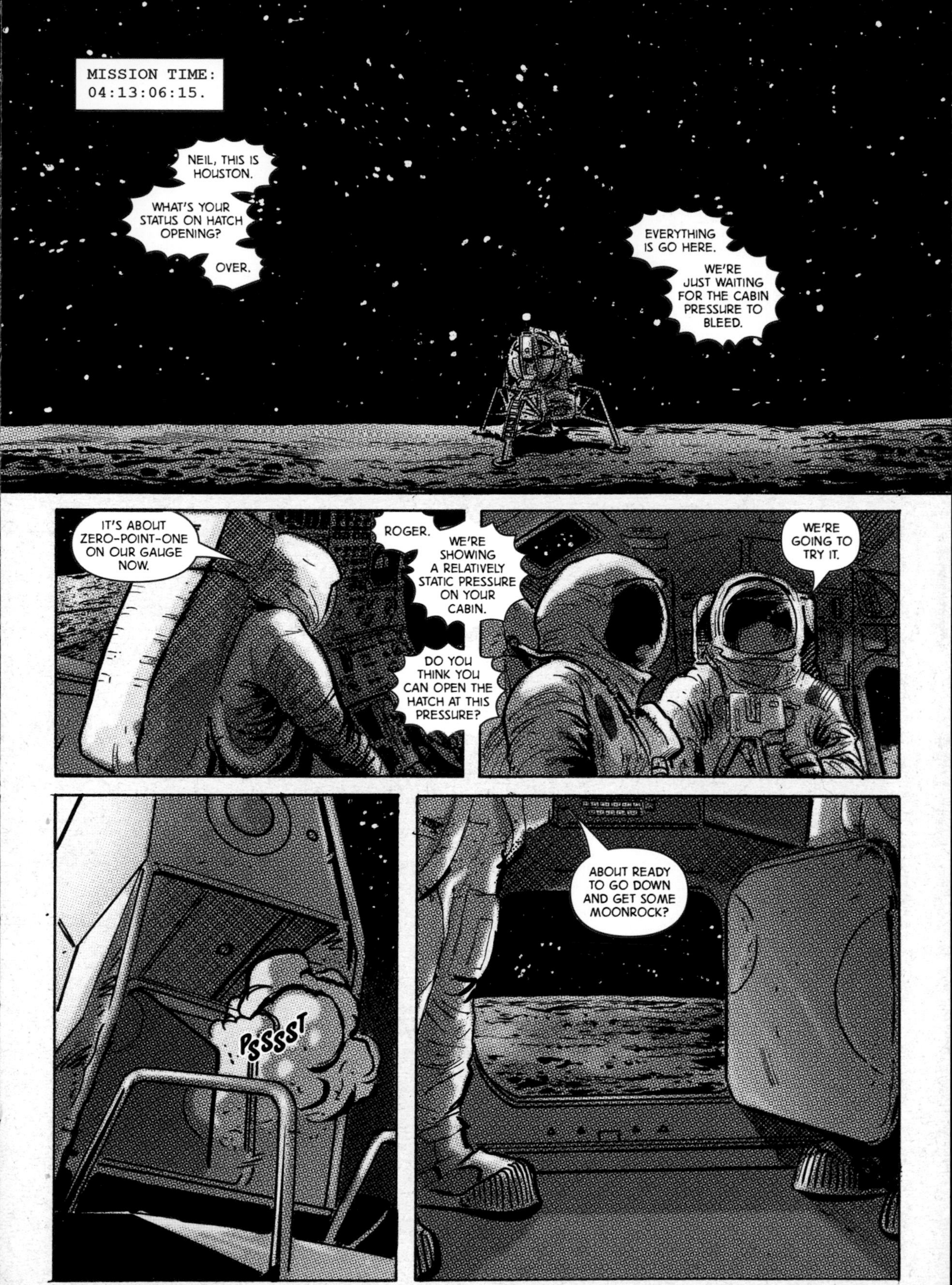
MISSION TIME: 04:13:06:15.
NEIL, THIS IS HOUSTON.
WHAT'S YOUR STATUS ON HATCH OPENING?
OVER.
EVERYTHING IS GO HERE.
WE'RE JUST WAITING FOR THE CABIN PRESSURE TO BLEED.
IT'S ABOUT ZERO-POINT-ONE ON OUR GAUGE NOW.
ROGER.
WE'RE SHOWING A RELATIVELY STATIC PRESSURE ON YOUR CABIN.
DO YOU THINK YOU CAN OPEN THE HATCH AT THIS PRESSURE?
WE'RE GOING TO TRY IT.
PSSSST
ABOUT READY TO GO DOWN AND GET SOME MOONROCK?

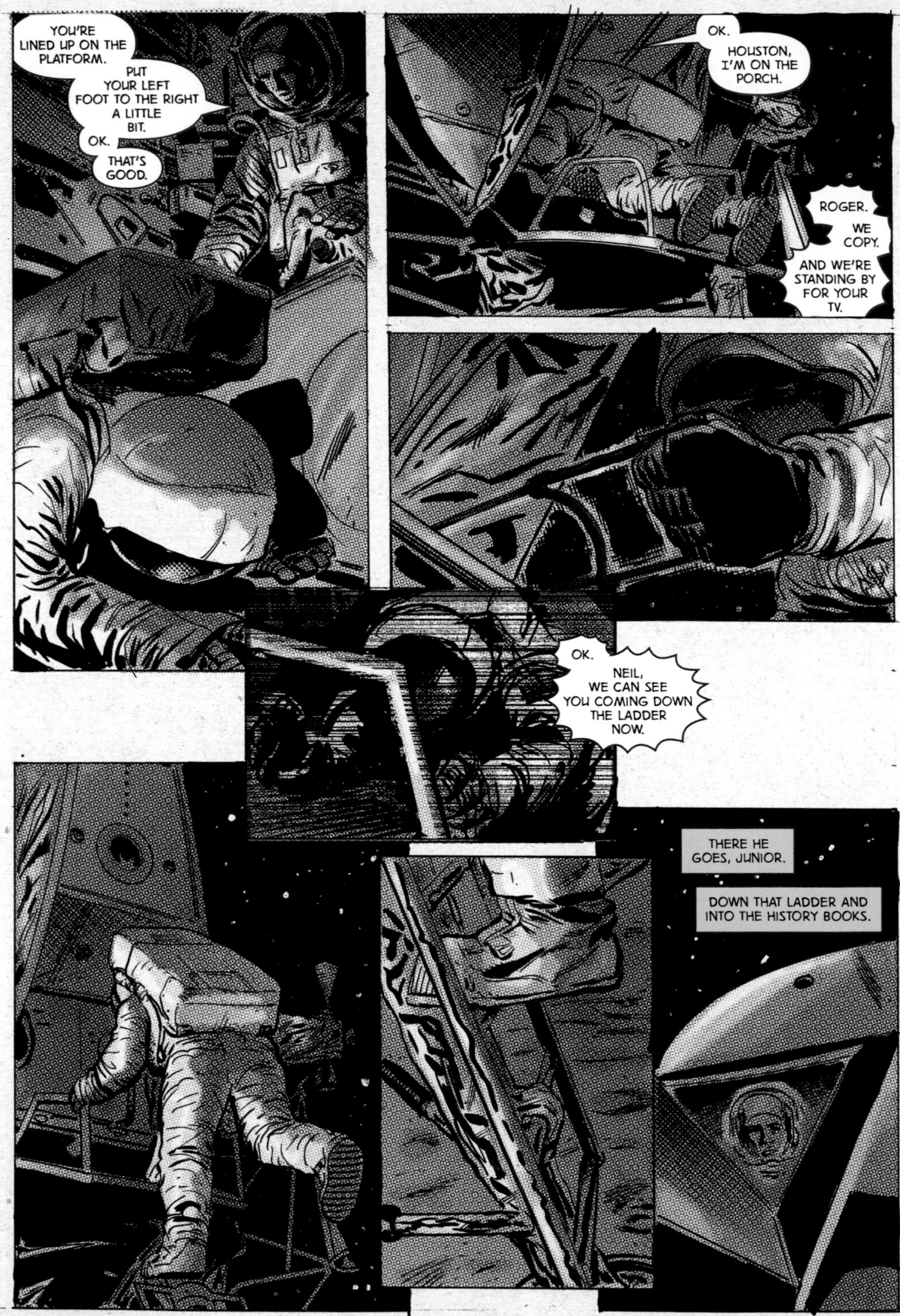
YOU'RE LINED UP ON THE PLATFORM.
PUT YOUR LEFT FOOT TO THE RIGHT A LITTLE BIT.
OK.
THAT'S GOOD.
OK.
HOUSTON, I'M ON THE PORCH.
ROGER.
WE COPY.
AND WE'RE STANDING BY FOR YOUR TV.
OK.
NEIL, WE CAN SEE YOU COMING DOWN THE LADDER NOW.
THERE HE GOES, JUNIOR.
DOWN THAT LADDER AND INTO THE HISTORY BOOKS.

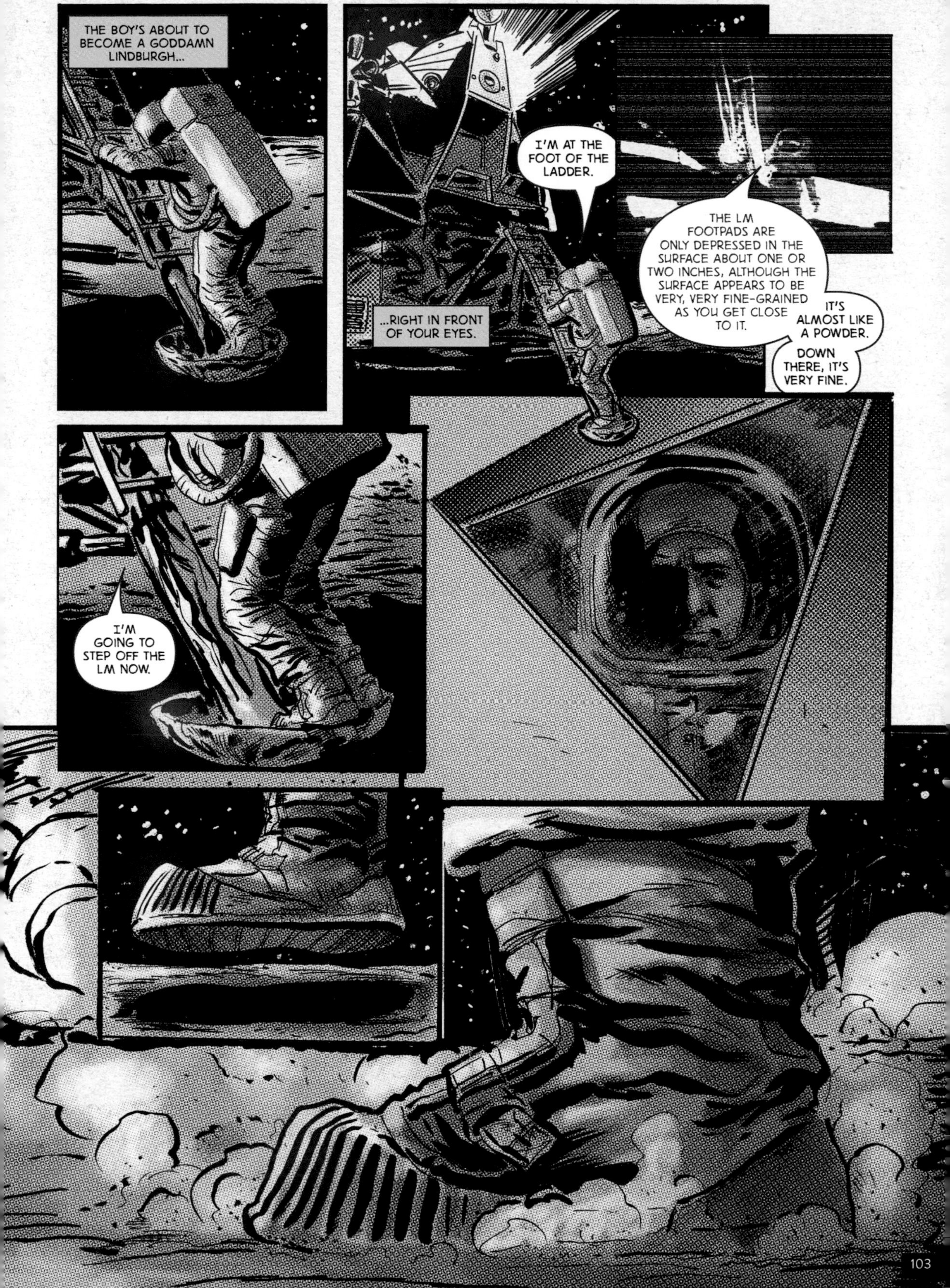
THE BOY'S ABOUT TO BECOME A GODDAMN LINDBURGH...
...RIGHT IN FRONT OF YOUR EYES.
I'M AT THE FOOT OF THE LADDER.
THE LM FOOTPADS ARE ONLY DEPRESSED IN THE SURFACE ABOUT ONE OR TWO INCHES, ALTHOUGH THE SURFACE APPEARS TO BE VERY, VERY FINE-GRAINED AS YOU GET CLOSE TO IT.
IT'S ALMOST LIKE A POWDER.
DOWN THERE, IT'S VERY FINE.
I'M GOING TO STEP OFF THE LM NOW.

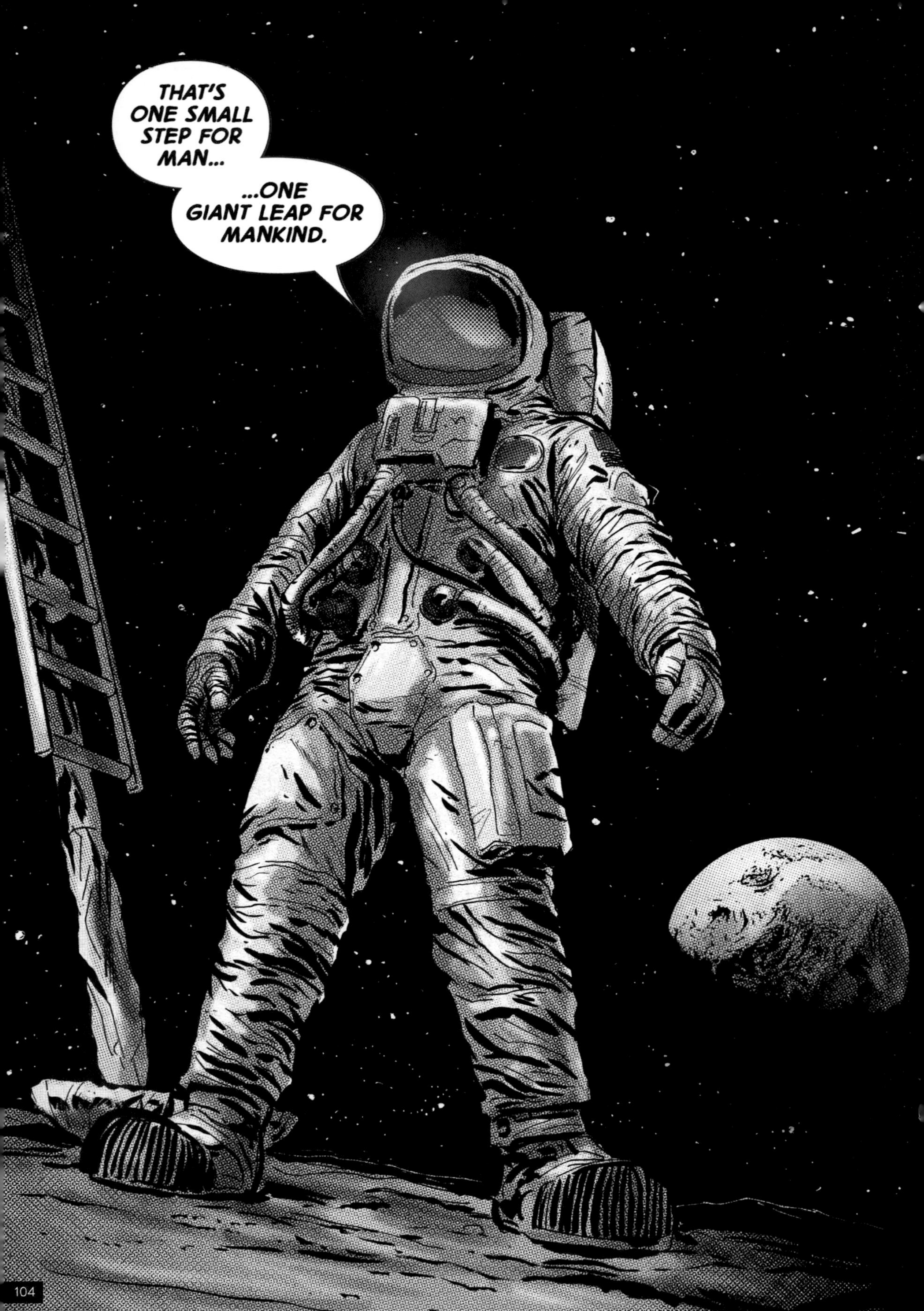
THAT'S ONE SMALL STEP FOR MAN...
...ONE GIANT LEAP FOR MANKIND.

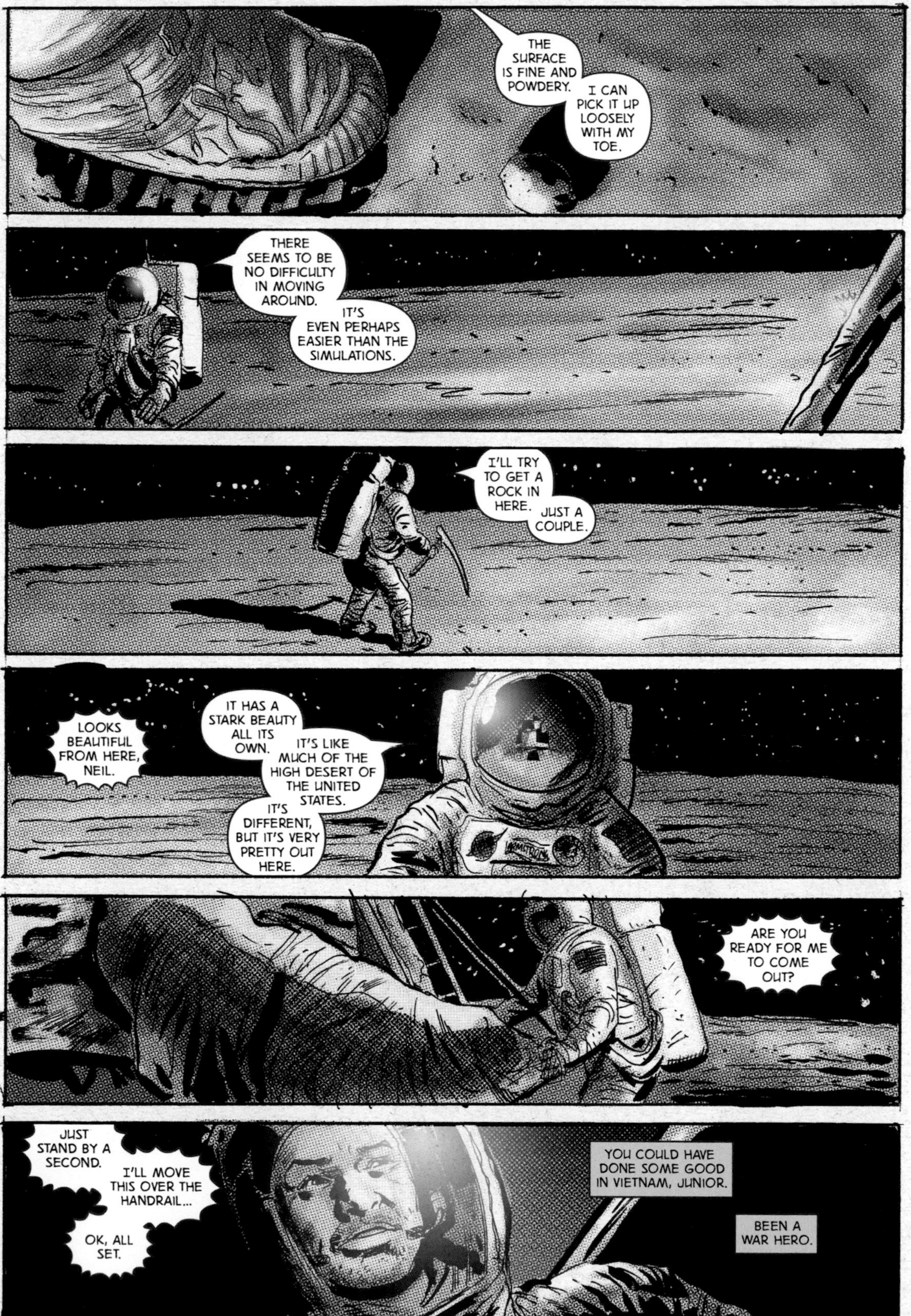
THE SURFACE IS FINE AND POWDERY.
I CAN PICK IT UP LOOSELY WITH MY TOE.
THERE SEEMS TO BE NO DIFFICULTY IN MOVING AROUND.
IT'S EVEN PERHAPS EASIER THAN THE SIMULATIONS.
I'LL TRY TO GET A ROCK IN HERE.
JUST A COUPLE.
LOOKS BEAUTIFUL FROM HERE, NEIL.
IT HAS A STARK BEAUTY ALL ITS OWN.
IT'S LIKE MUCH OF THE HIGH DESERT OF THE UNITED STATES.
IT'S DIFFERENT, BUT IT'S VERY PRETTY OUT HERE.
ARE YOU READY FOR ME TO COME OUT?
JUST STAND BY A SECOND.
I'LL MOVE THIS OVER THE HANDRAIL...
OK, ALL SET.
YOU COULD HAVE DONE SOME GOOD IN VIETNAM, JUNIOR.
BEEN A WAR HERO.

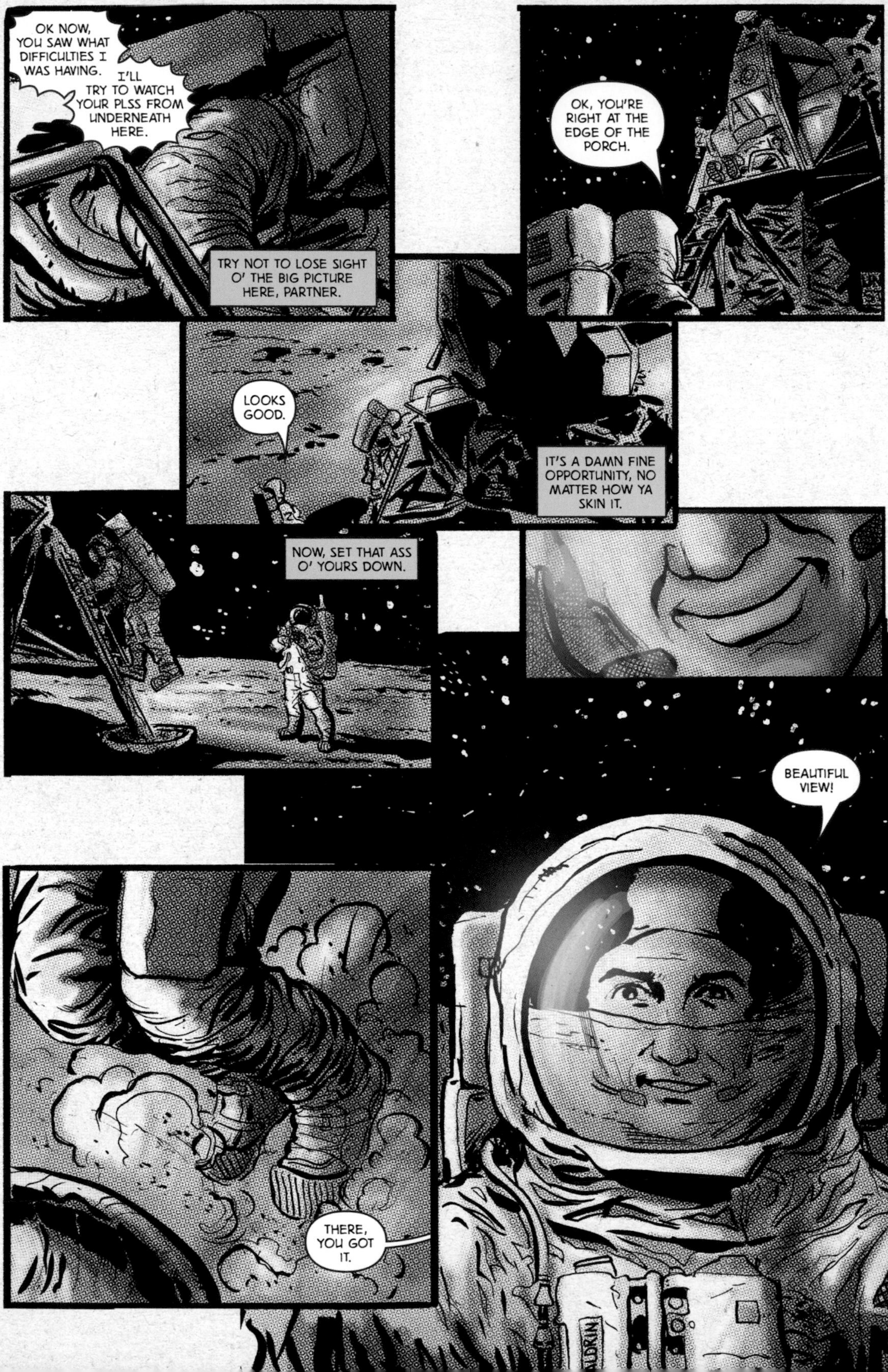
OK NOW, YOU SAW WHAT DIFFICULTIES I WAS HAVING.
I'LL TRY TO WATCH YOUR PLSS FROM UNDERNEATH HERE.
TRY NOT TO LOSE SIGHT O' THE BIG PICTURE HERE, PARTNER.
OK, YOU'RE RIGHT AT THE EDGE OF THE PORCH.
LOOKS GOOD.
IT'S A DAMN FINE OPPORTUNITY, NO MATTER HOW YA SKIN IT.
NOW, SET THAT ASS O' YOURS DOWN.
THERE, YOU GOT IT.
BEAUTIFUL VIEW!
ALDRIN

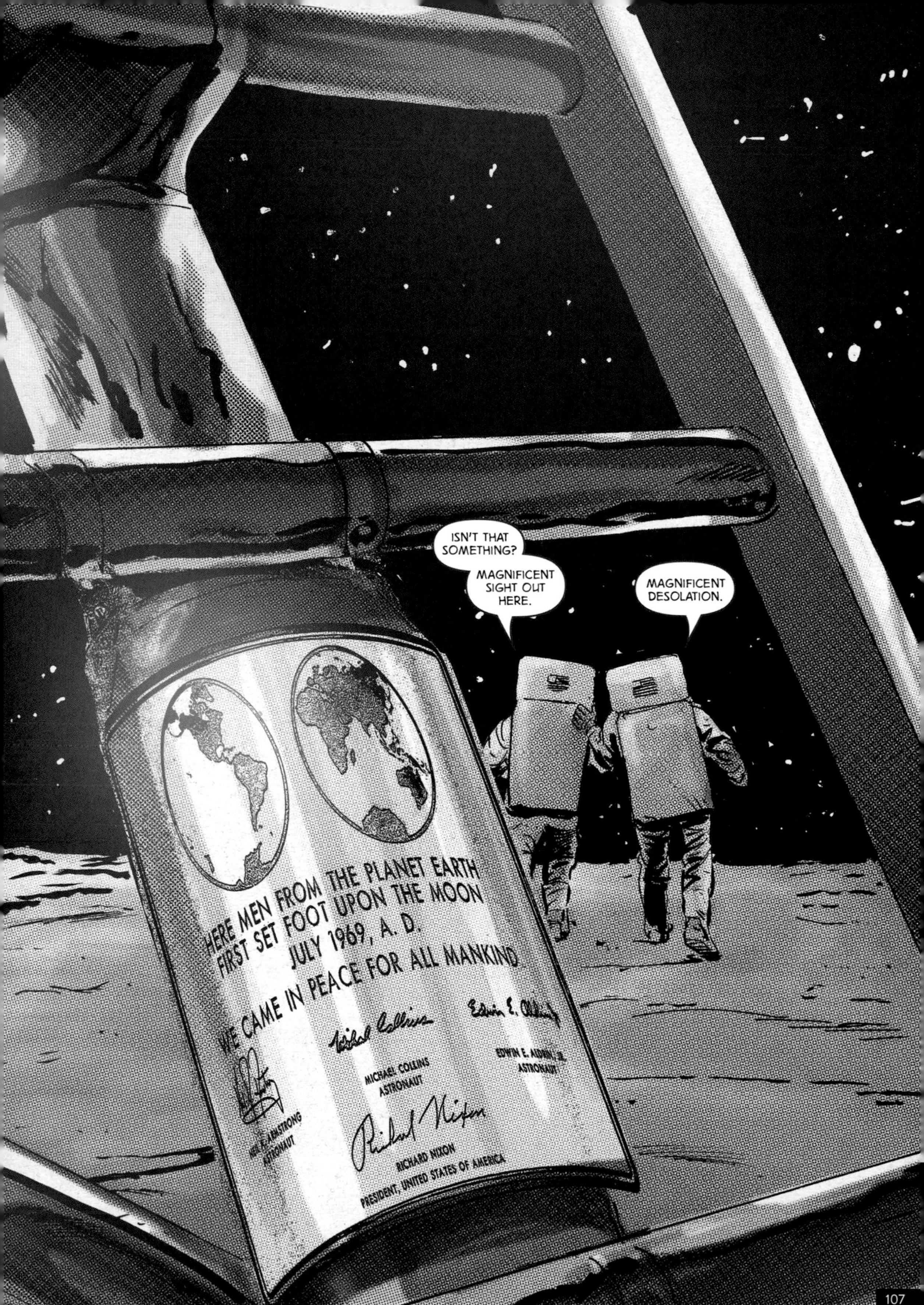

ISN'T THAT SOMETHING?
MAGNIFICENT SIGHT OUT HERE.
MAGNIFICENT DESOLATION.
HERE MEN FROM THE PLANET EARTH
FIRST SET FOOT UPON THE MOON
JULY 1969, A. D.
WE CAME IN PEACE FOR ALL MANKIND
MICHAEL COLLINS
ASTRONAUT
EDWIN E. ALDRIN, JR.
ASTRONAUT
RICHARD NIXON
PRESIDENT, UNITED STATES OF AMERICA

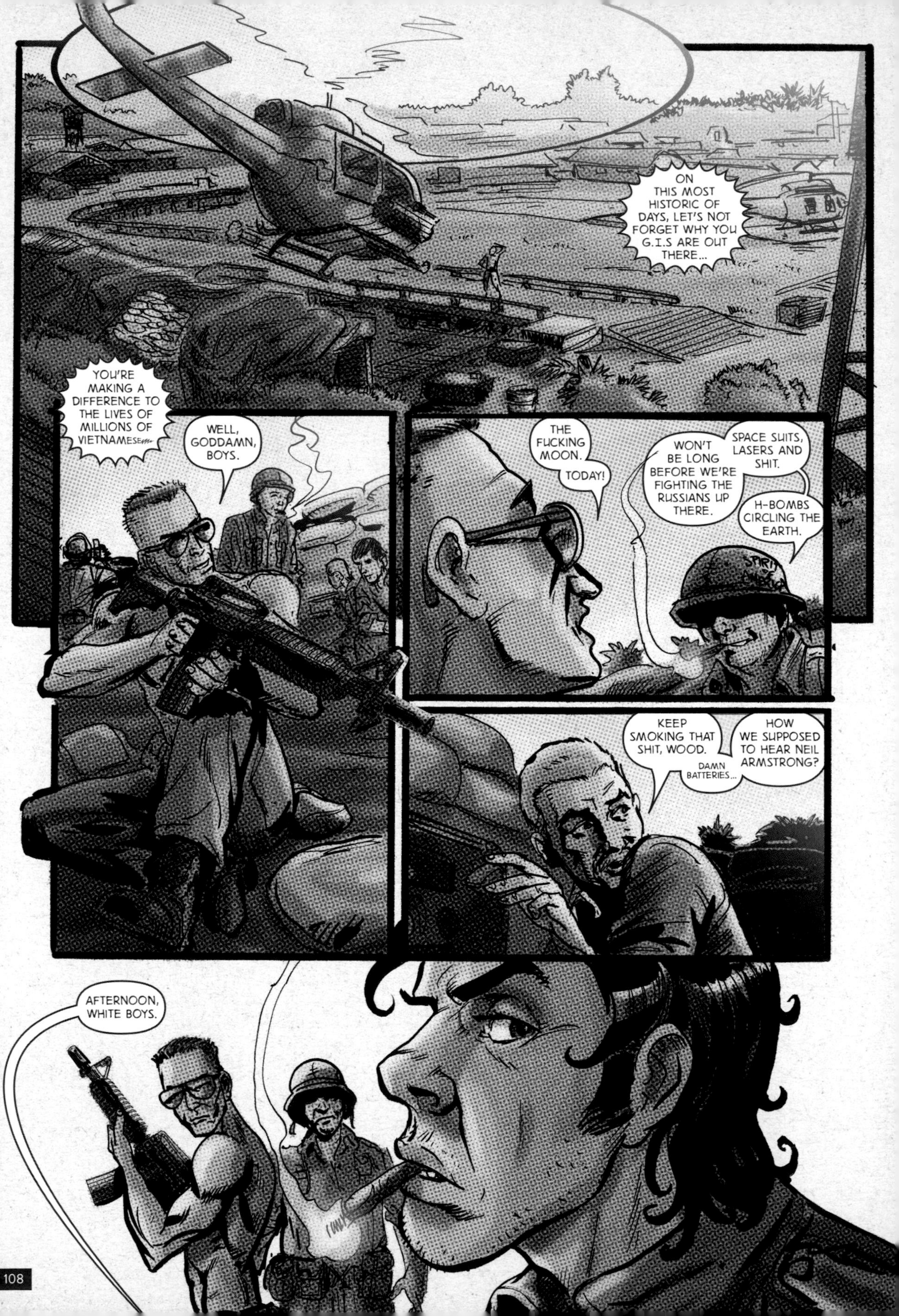
ON THIS MOST HISTORIC OF DAYS, LET'S NOT FORGET WHY YOU G.I.S ARE OUT THERE...
YOU'RE MAKING A DIFFERENCE TO THE LIVES OF MILLIONS OF VIETNAMESE
WELL, GODDAMN, BOYS.
THE FUCKING MOON.
TODAY!
WON'T BE LONG BEFORE WE'RE FIGHTING THE RUSSIANS UP THERE.
SPACE SUITS, LASERS AND SHIT.
H-BOMBS CIRCLING THE EARTH.
KEEP SMOKING THAT SHIT, WOOD.
HOW WE SUPPOSED TO HEAR NEIL ARMSTRONG?
DAMN BATTERIES...
AFTERNOON, WHITE BOYS.

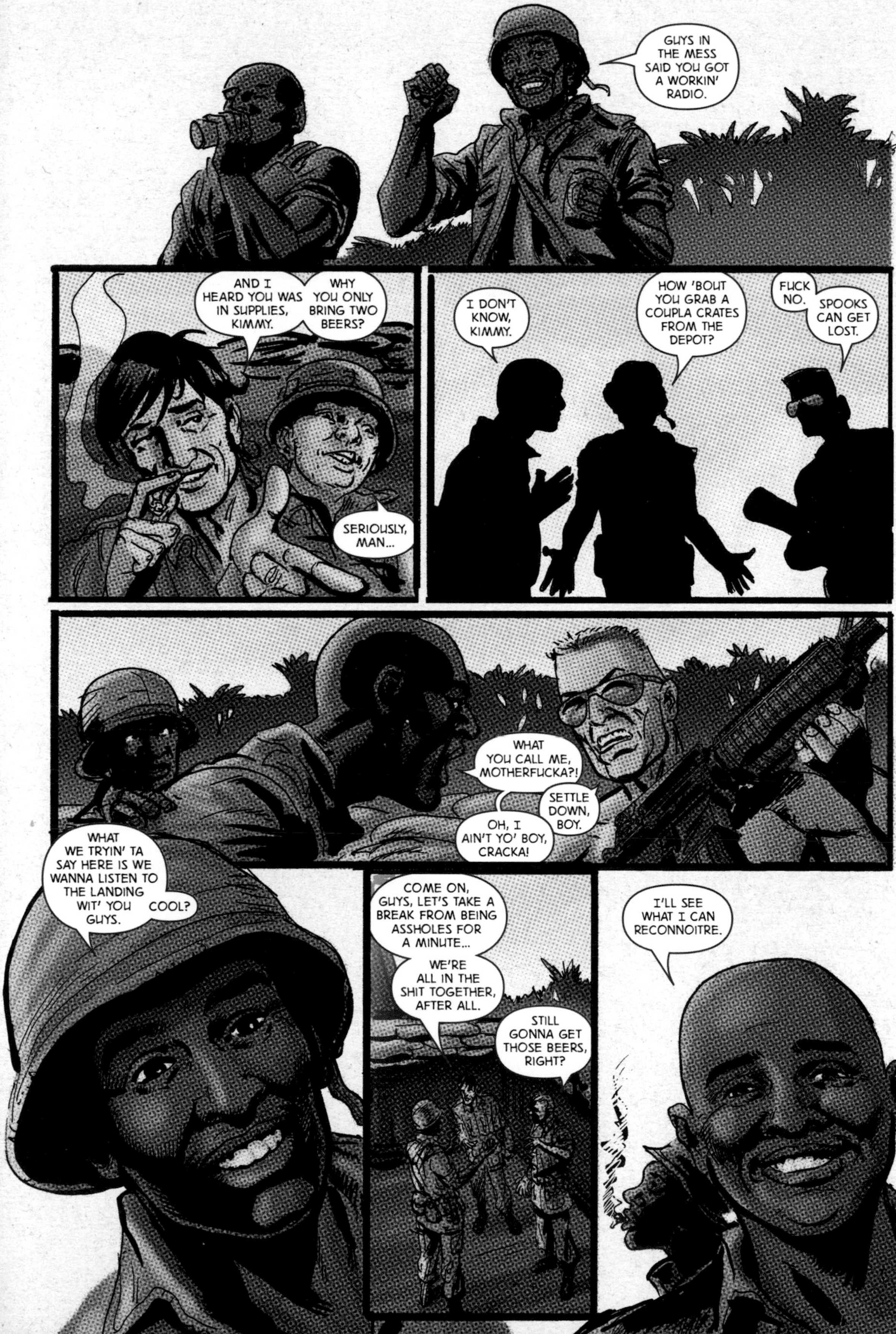

GUYS IN THE MESS SAID YOU GOT A WORKIN' RADIO.
AND I HEARD YOU WAS IN SUPPLIES, KIMMY.
WHY YOU ONLY BRING TWO BEERS?
SERIOUSLY, MAN...
I DON'T KNOW, KIMMY.
HOW 'BOUT YOU GRAB A COUPLA CRATES FROM THE DEPOT?
FUCK NO.
SPOOKS CAN GET LOST.
WHAT YOU CALL ME, MOTHERFUCKA?!
SETTLE DOWN, BOY.
OH, I AIN'T YO' BOY, CRACKA!
WHAT WE TRYIN' TA SAY HERE IS WE WANNA LISTEN TO THE LANDING WIT' YOU GUYS.
COOL?
COME ON, GUYS, LET'S TAKE A BREAK FROM BEING ASSHOLES FOR A MINUTE...
WE'RE ALL IN THE SHIT TOGETHER, AFTER ALL.
STILL GONNA GET THOSE BEERS, RIGHT?
I'LL SEE WHAT I CAN RECONNOITRE.

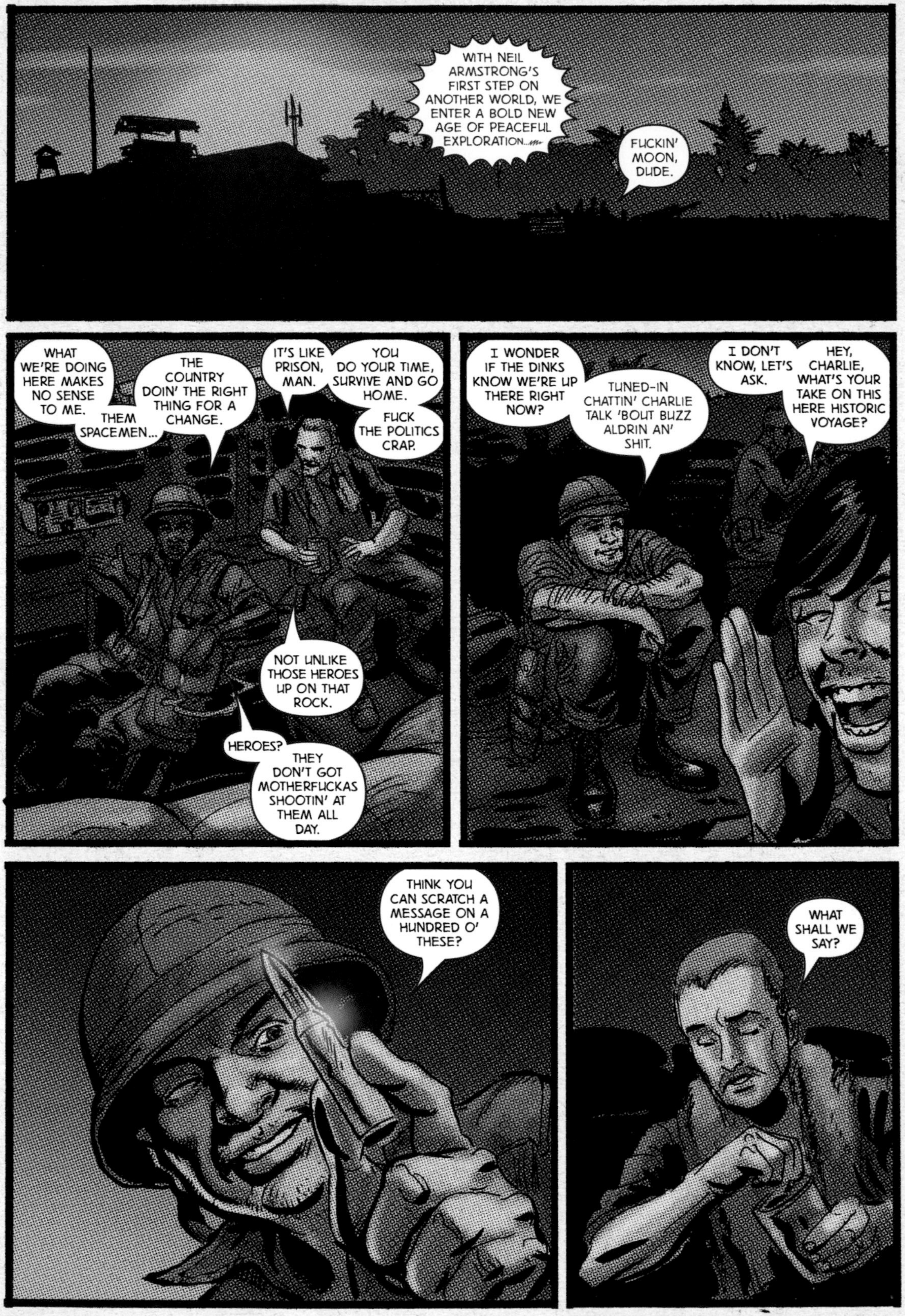
WITH NEIL ARMSTRONG'S FIRST STEP ON ANOTHER WORLD, WE ENTER A BOLD NEW AGE OF PEACEFUL EXPLORATION...
FUCKIN' MOON, DUDE.
WHAT WE'RE DOING HERE MAKES NO SENSE TO ME.
THEM SPACEMEN...
THE COUNTRY DOIN' THE RIGHT THING FOR A CHANGE.
IT'S LIKE PRISON, MAN.
YOU DO YOUR TIME, SURVIVE AND GO HOME.
FUCK THE POLITICS CRAP.
NOT UNLIKE THOSE HEROES UP ON THAT ROCK.
HEROES?
THEY DON'T GOT MOTHERFUCKAS SHOOTIN' AT THEM ALL DAY.
I WONDER IF THE DINKS KNOW WE'RE UP THERE RIGHT NOW?
TUNED-IN CHATTIN' CHARLIE TALK 'BOUT BUZZ ALDRIN AN' SHIT.
I DON'T KNOW, LET'S ASK.
HEY, CHARLIE, WHAT'S YOUR TAKE ON THIS HERE HISTORIC VOYAGE?
THINK YOU CAN SCRATCH A MESSAGE ON A HUNDRED O' THESE?
WHAT SHALL WE SAY?

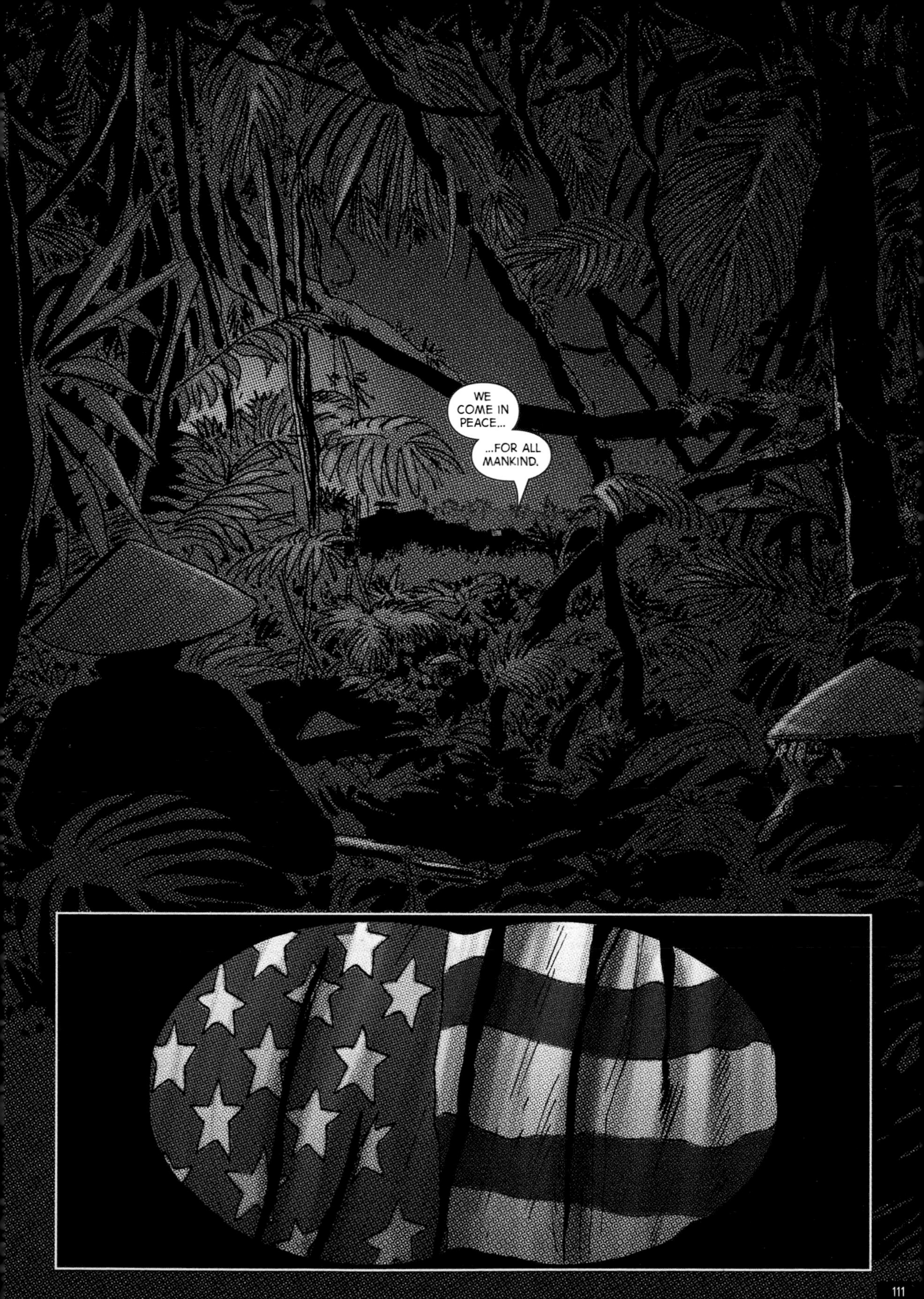
WE COME IN PEACE...
...FOR ALL MANKIND.

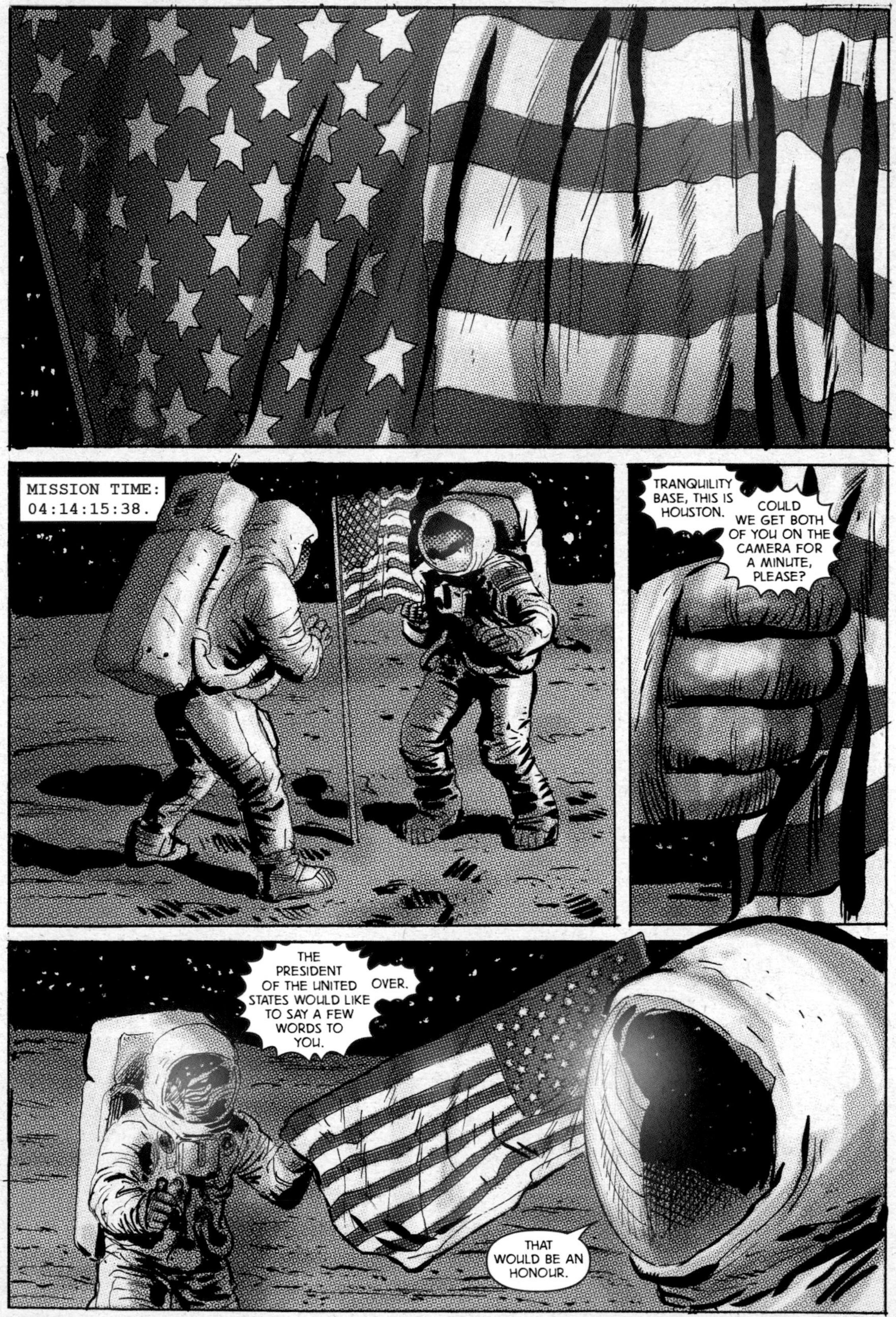
MISSION TIME: 04:14:15:38.
TRANQUILITY BASE, THIS IS HOUSTON.
COULD WE GET BOTH OF YOU ON THE CAMERA FOR A MINUTE, PLEASE?
THE PRESIDENT OF THE UNITED STATES WOULD LIKE TO SAY A FEW WORDS TO YOU.
OVER.
THAT WOULD BE AN HONOUR.

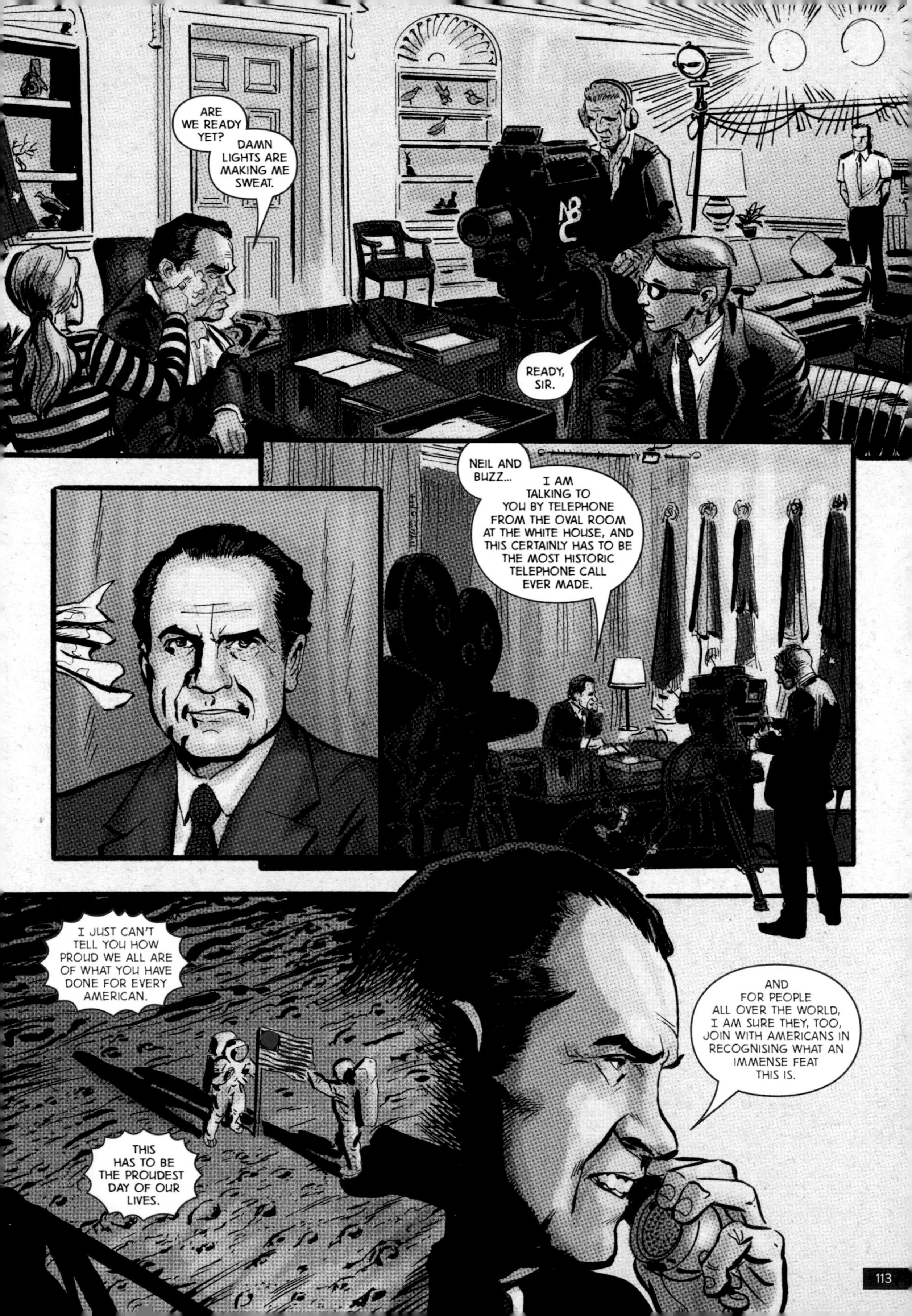
ARE WE READY YET?
DAMN LIGHTS ARE MAKING ME SWEAT.
READY, SIR.
NEIL AND BUZZ...
I AM TALKING TO YOU BY TELEPHONE FROM THE OVAL ROOM AT THE WHITE HOUSE, AND THIS CERTAINLY HAS TO BE THE MOST HISTORIC TELEPHONE CALL EVER MADE.
I JUST CAN'T TELL YOU HOW PROUD WE ALL ARE OF WHAT YOU HAVE DONE FOR EVERY AMERICAN.
THIS HAS TO BE THE PROUDEST DAY OF OUR LIVES.
AND FOR PEOPLE ALL OVER THE WORLD, I AM SURE THEY, TOO, JOIN WITH AMERICANS IN RECOGNISING WHAT AN IMMENSE FEAT THIS IS.

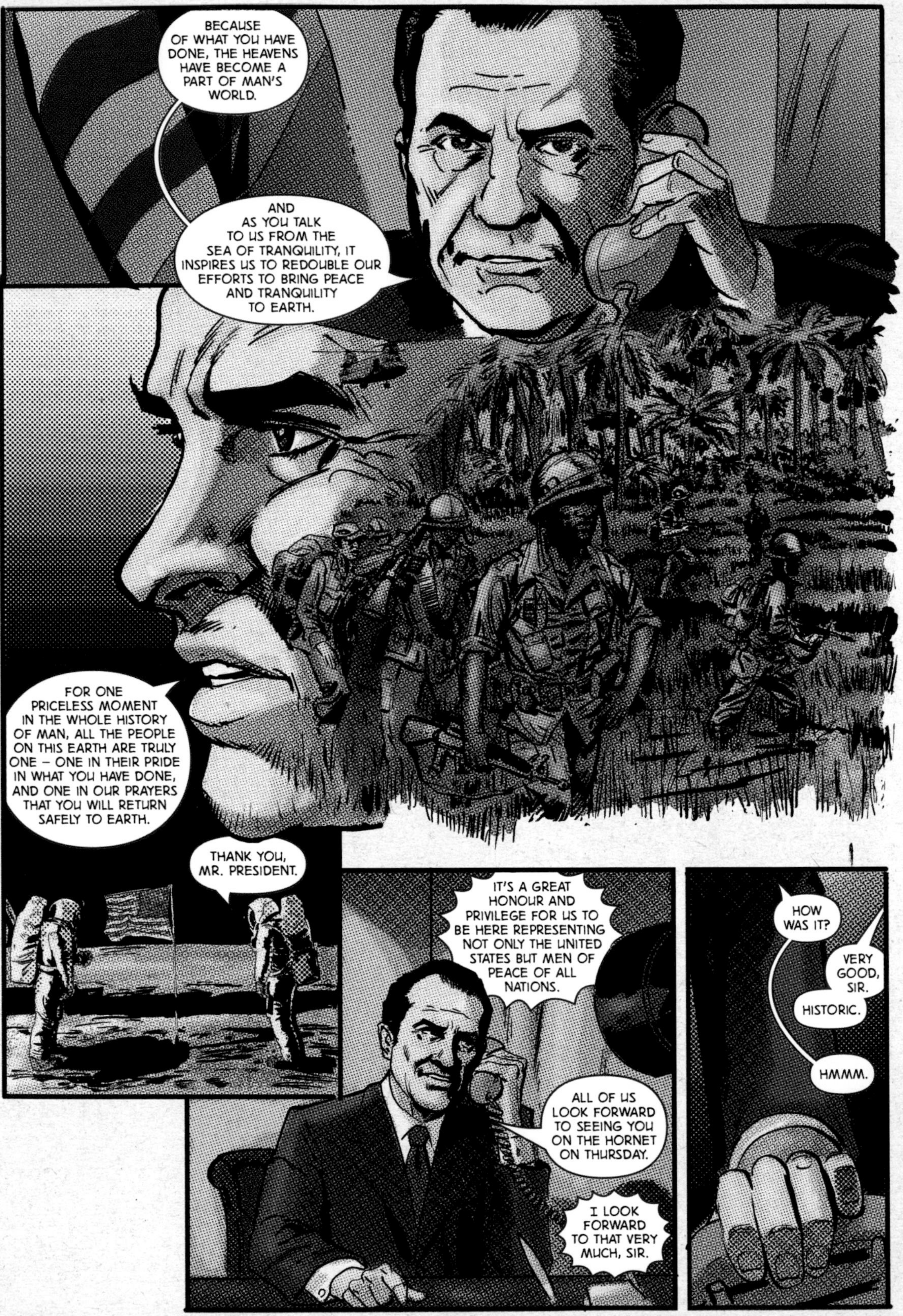
BECAUSE OF WHAT YOU HAVE DONE, THE HEAVENS HAVE BECOME A PART OF MAN'S WORLD.
AND AS YOU TALK TO US FROM THE SEA OF TRANQUILITY, IT INSPIRES US TO REDOUBLE OUR EFFORTS TO BRING PEACE AND TRANQUILITY TO EARTH.
FOR ONE PRICELESS MOMENT IN THE WHOLE HISTORY OF MAN, ALL THE PEOPLE ON THIS EARTH ARE TRULY ONE – ONE IN THEIR PRIDE IN WHAT YOU HAVE DONE, AND ONE IN OUR PRAYERS THAT YOU WILL RETURN SAFELY TO EARTH.
THANK YOU, MR. PRESIDENT.
IT'S A GREAT HONOUR AND PRIVILEGE FOR US TO BE HERE REPRESENTING NOT ONLY THE UNITED STATES BUT MEN OF PEACE OF ALL NATIONS.
ALL OF US LOOK FORWARD TO SEEING YOU ON THE HORNET ON THURSDAY.
I LOOK FORWARD TO THAT VERY MUCH, SIR.
HOW WAS IT?
VERY GOOD, SIR.
HISTORIC.
HMMM.

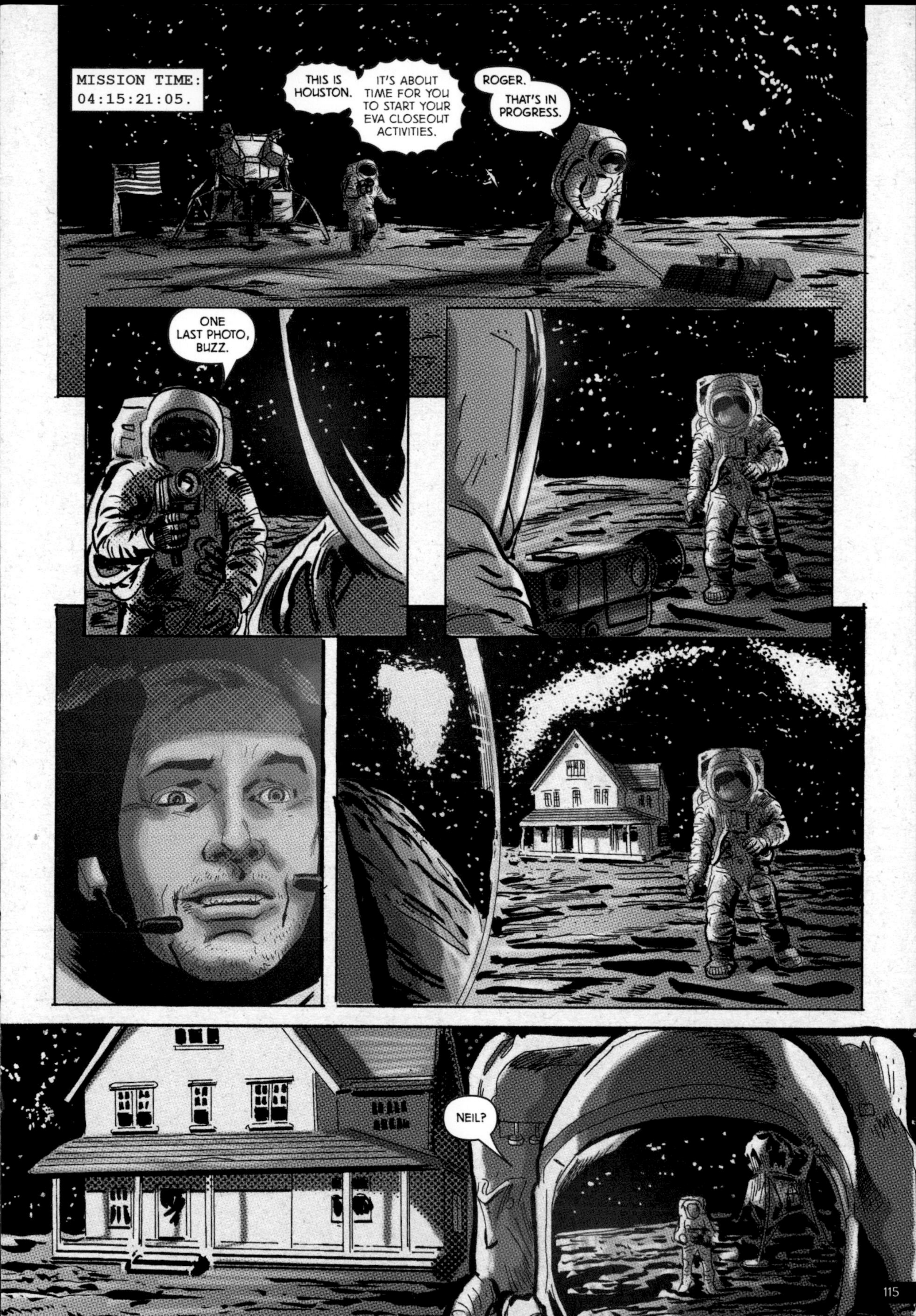
MISSION TIME:
04:15:21:05.
THIS IS HOUSTON.
IT'S ABOUT TIME FOR YOU TO START YOUR EVA CLOSEOUT ACTIVITIES.
ROGER.
THAT'S IN PROGRESS.
ONE LAST PHOTO, BUZZ.
NEIL?

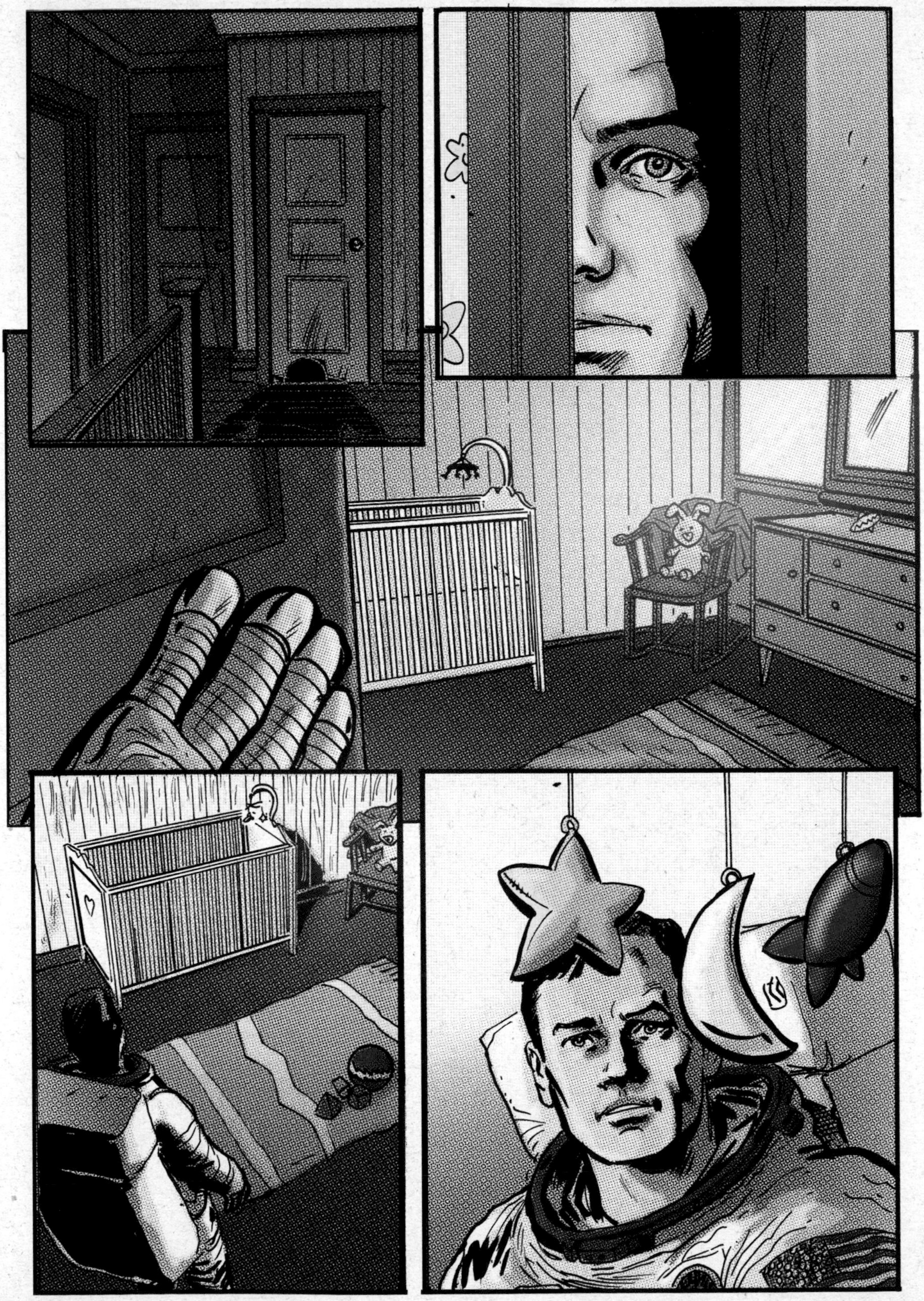

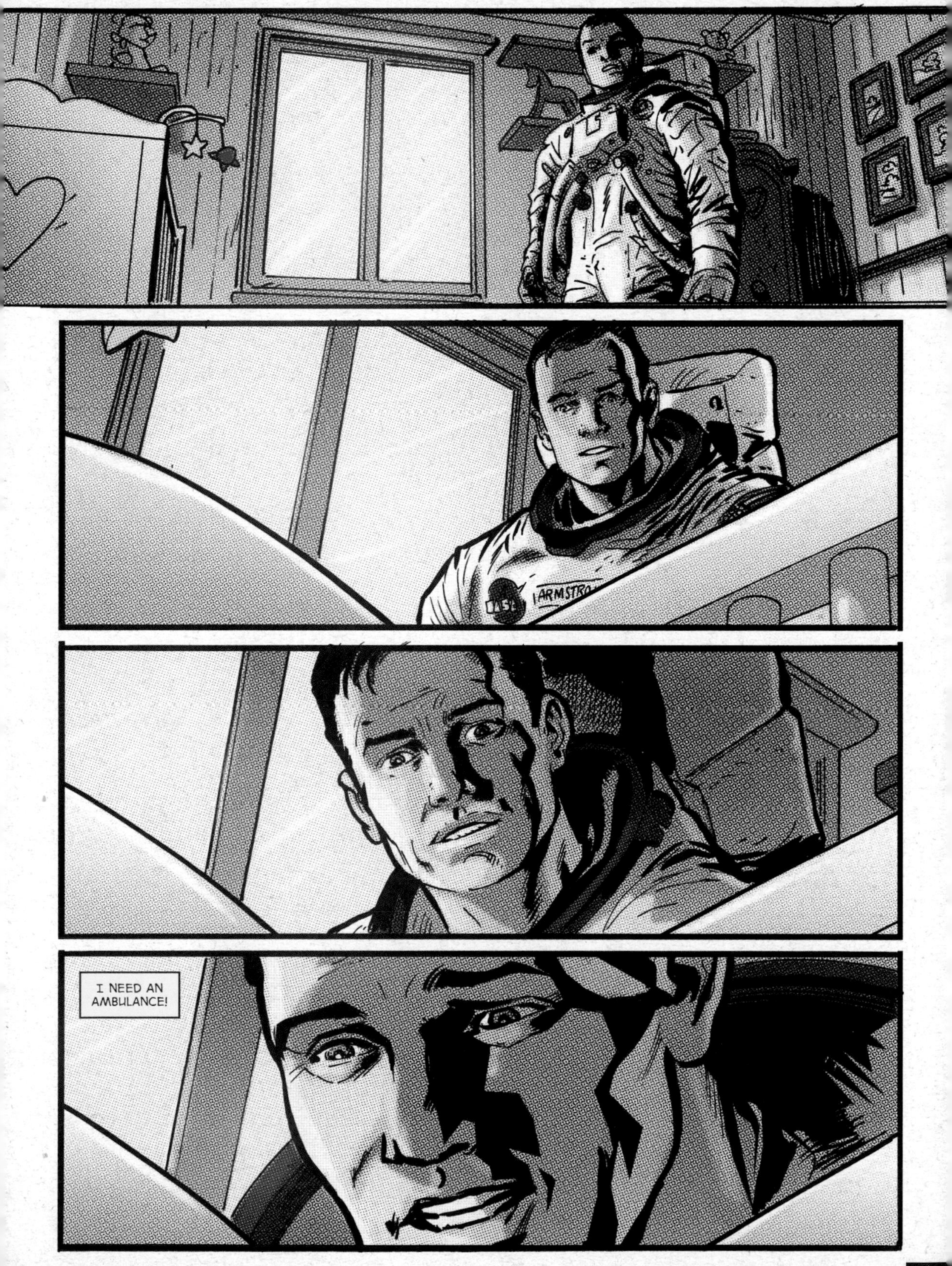
ARMSTRO
I NEED AN AMBULANCE!

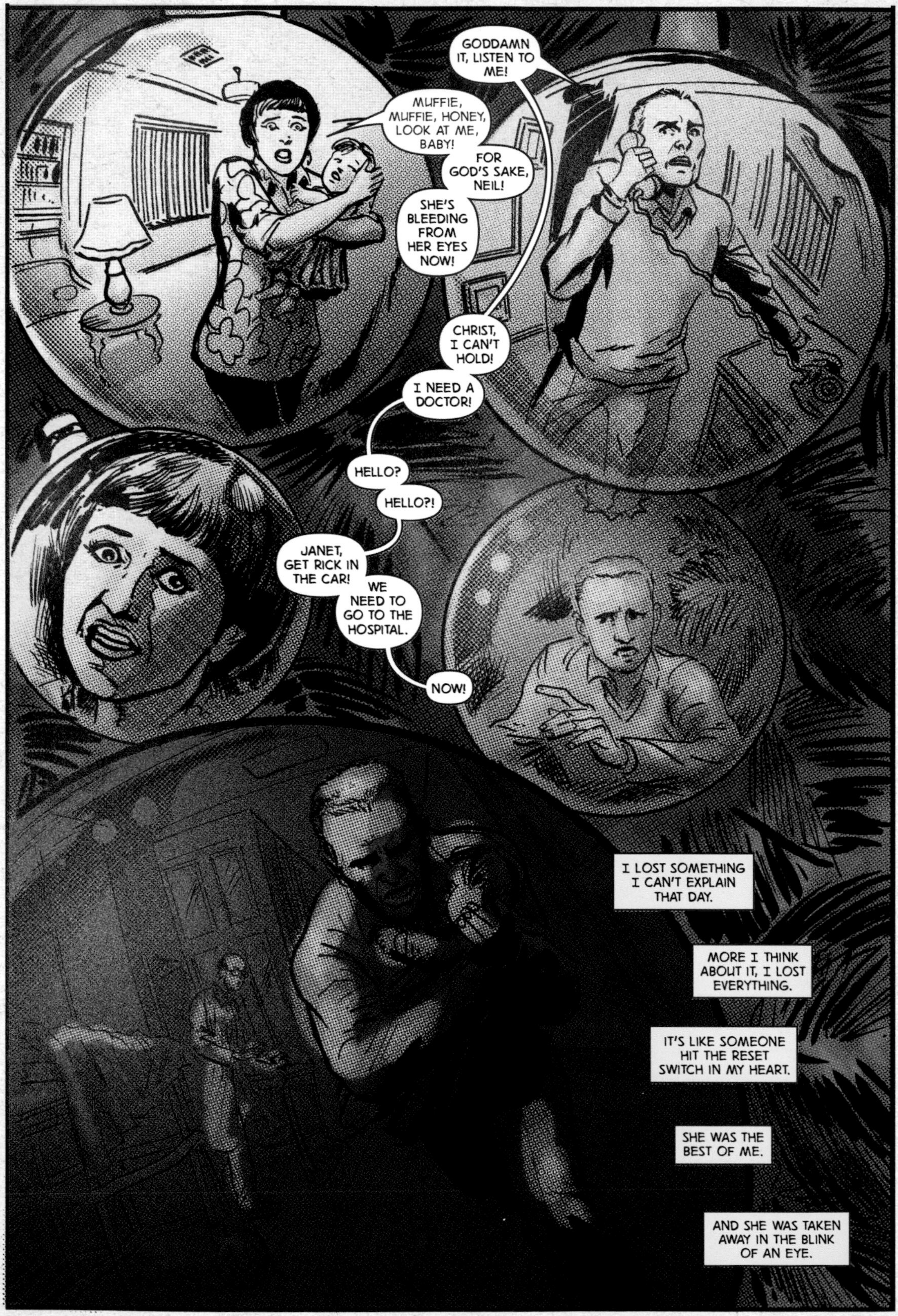
MUFFIE, MUFFIE, HONEY, LOOK AT ME, BABY!
GODDAMN IT, LISTEN TO ME!
FOR GOD'S SAKE, NEIL!
SHE'S BLEEDING FROM HER EYES NOW!
CHRIST, I CAN'T HOLD!
I NEED A DOCTOR!
HELLO?
HELLO?!
JANET, GET RICK IN THE CAR!
WE NEED TO GO TO THE HOSPITAL.
NOW!
I LOST SOMETHING I CAN'T EXPLAIN THAT DAY.
MORE I THINK ABOUT IT, I LOST EVERYTHING.
IT'S LIKE SOMEONE HIT THE RESET SWITCH IN MY HEART.
SHE WAS THE BEST OF ME.
AND SHE WAS TAKEN AWAY IN THE BLINK OF AN EYE.

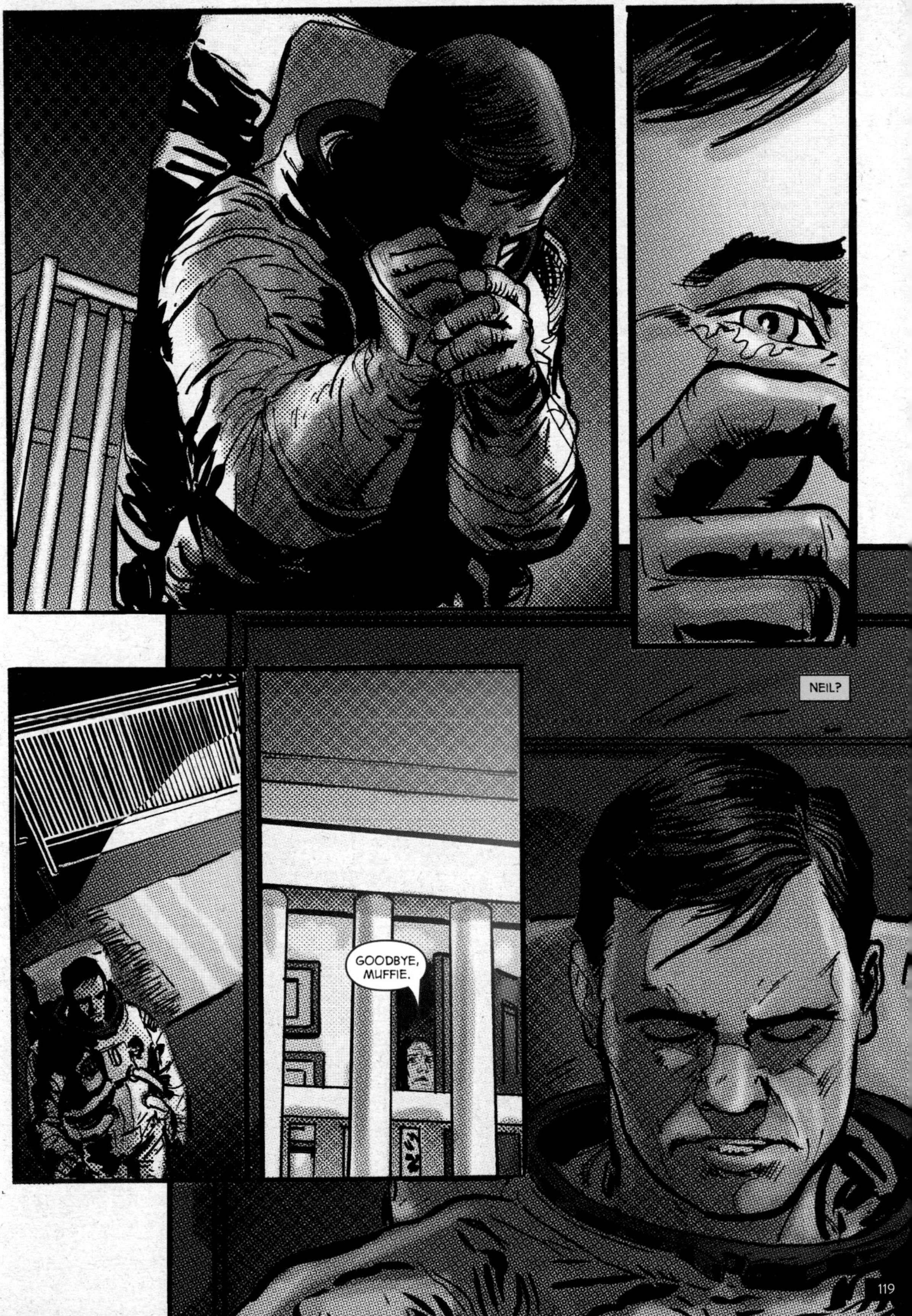
NEIL?
GOODBYE, MUFFIE.

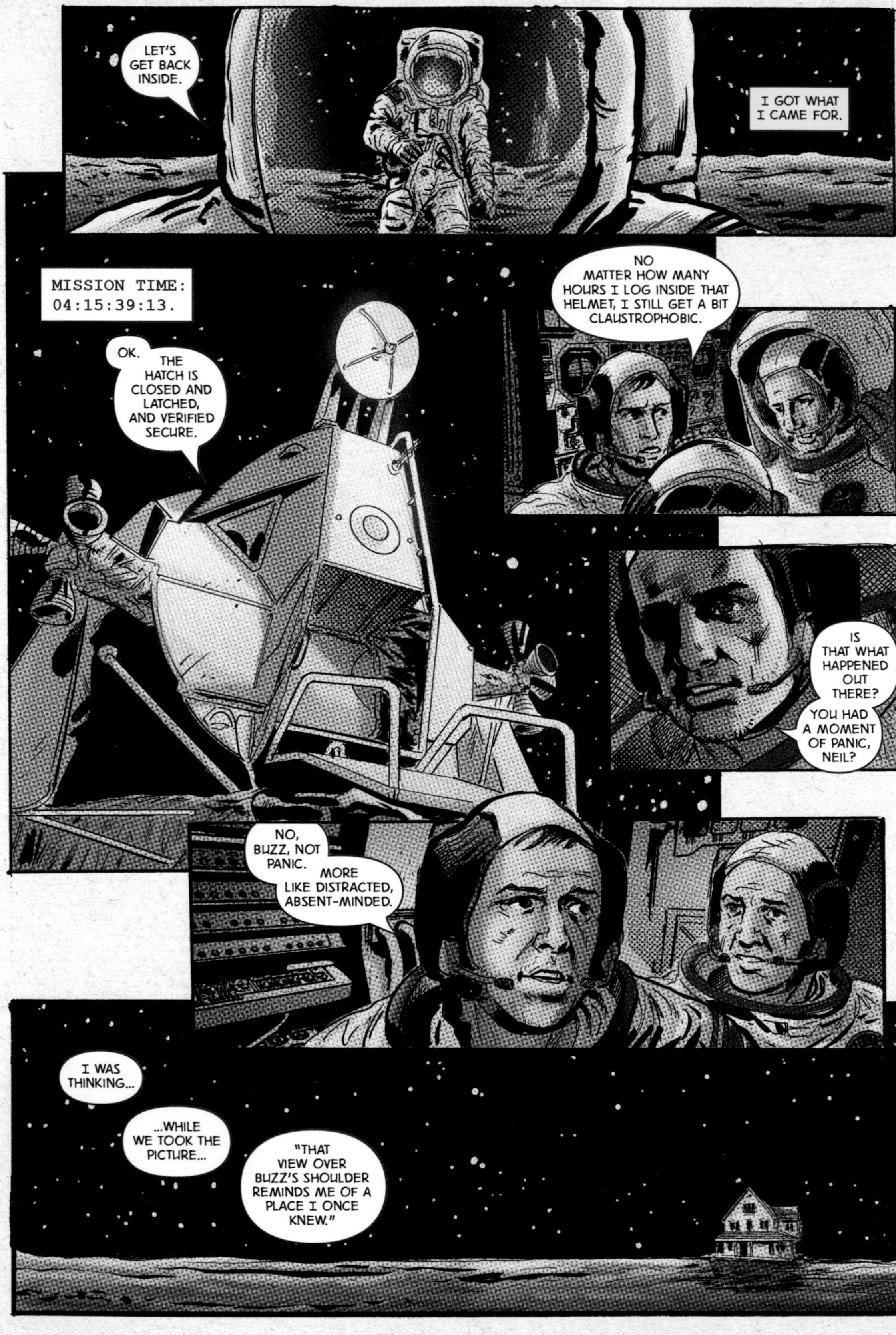
LET'S GET BACK INSIDE.
I GOT WHAT I CAME FOR.
MISSION TIME: 04:15:39:13.
OK. THE HATCH IS CLOSED AND LATCHED, AND VERIFIED SECURE.
NO MATTER HOW MANY HOURS I LOG INSIDE THAT HELMET, I STILL GET A BIT CLAUSTROPHOBIC.
IS THAT WHAT HAPPENED OUT THERE?
YOU HAD A MOMENT OF PANIC, NEIL?
NO, BUZZ, NOT PANIC.
MORE LIKE DISTRACTED, ABSENT-MINDED.
I WAS THINKING...
...WHILE WE TOOK THE PICTURE...
"THAT VIEW OVER BUZZ'S SHOULDER REMINDS ME OF A PLACE I ONCE KNEW."

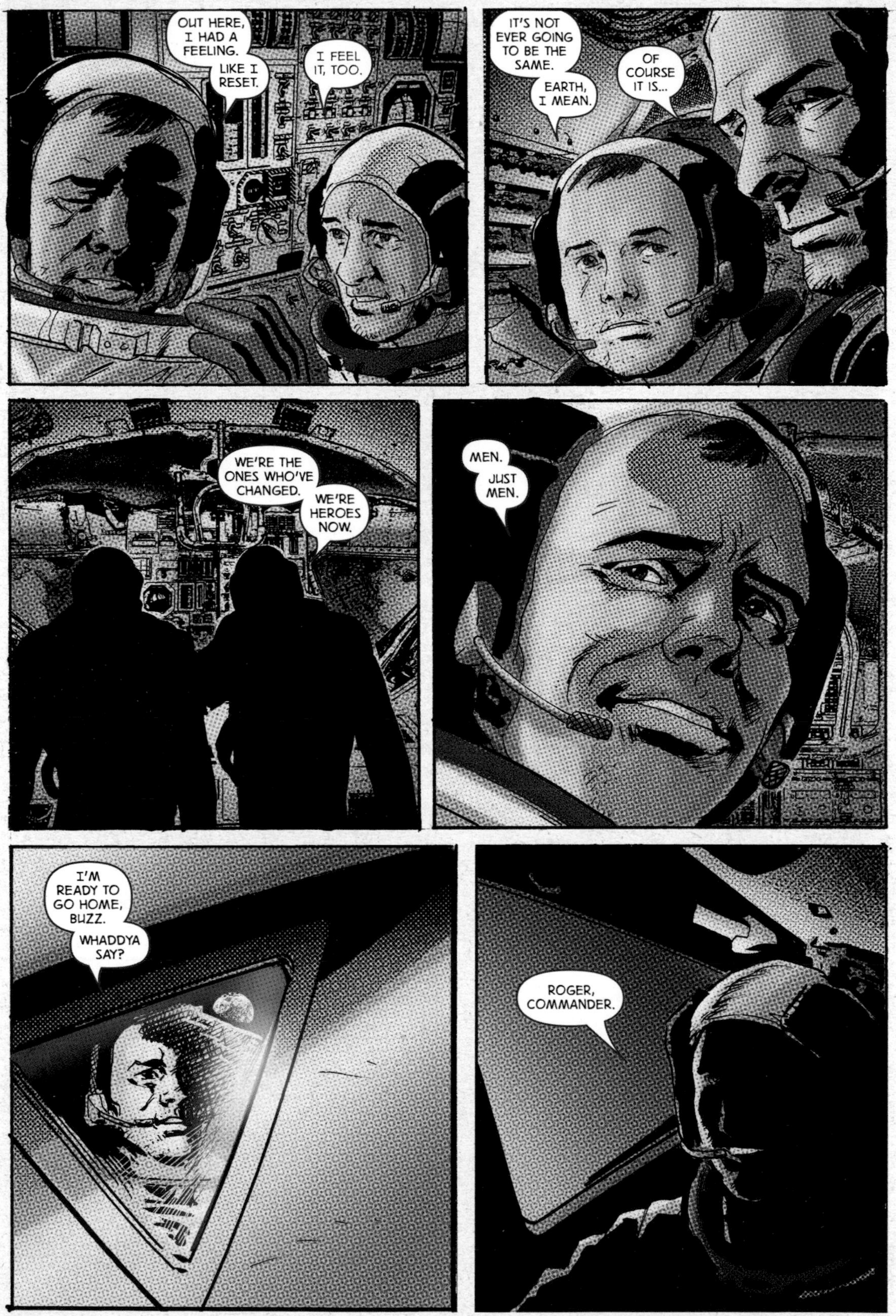
OUT HERE, I HAD A FEELING.
LIKE I RESET.
I FEEL IT, TOO.
IT'S NOT EVER GOING TO BE THE SAME.
EARTH, I MEAN.
OF COURSE IT IS...
WE'RE THE ONES WHO'VE CHANGED.
WE'RE HEROES NOW.
MEN. JUST MEN.
I'M READY TO GO HOME, BUZZ.
WHADDYA SAY?
ROGER, COMMANDER.

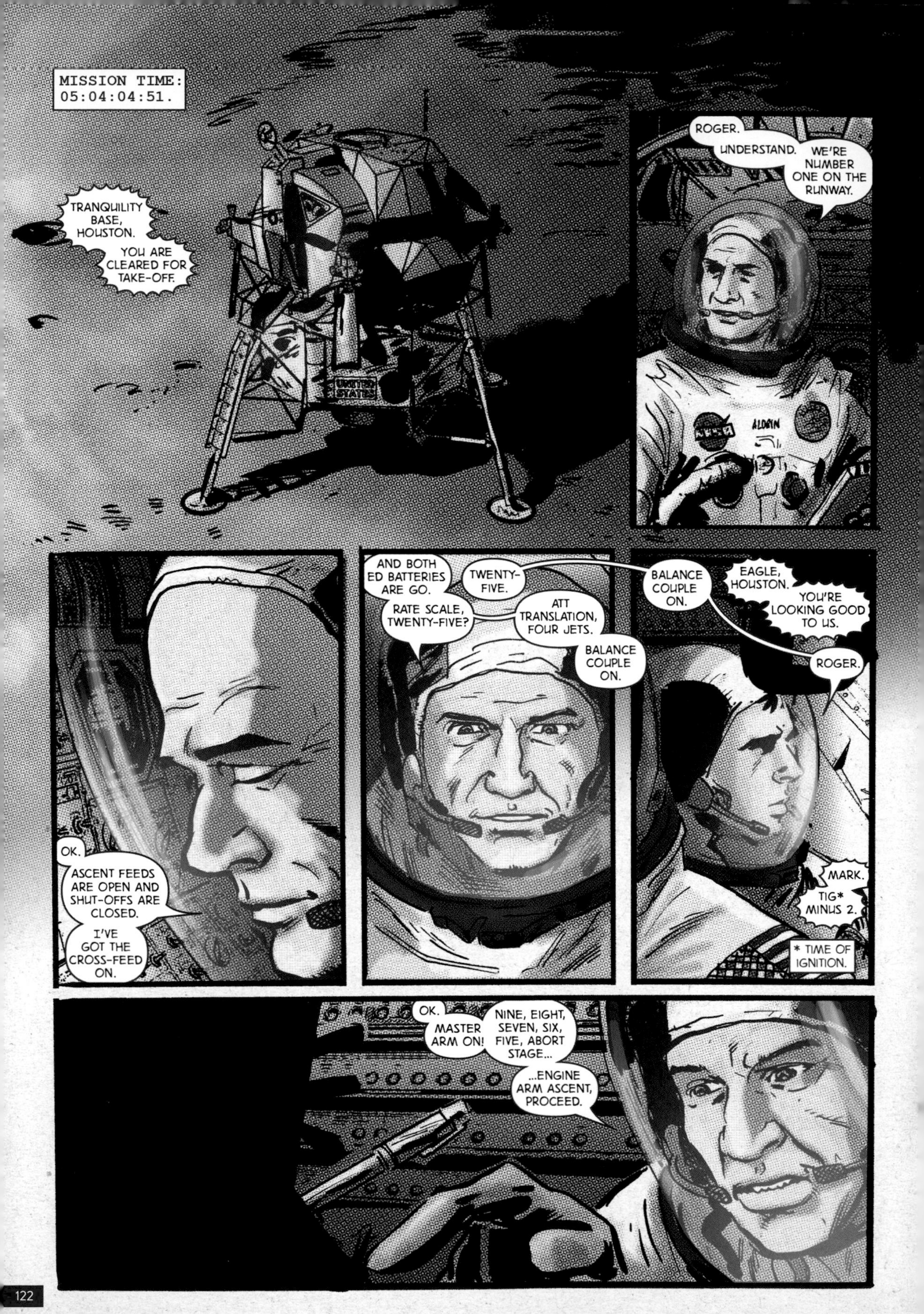
MISSION TIME:
05:04:04:51.
TRANQUILITY BASE, HOUSTON.
YOU ARE CLEARED FOR TAKE-OFF.
UNITED STATES
ROGER.
UNDERSTAND.
WE'RE NUMBER ONE ON THE RUNWAY.
ALDRIN
OK.
ASCENT FEEDS ARE OPEN AND SHUT-OFFS ARE CLOSED.
I'VE GOT THE CROSS-FEED ON.
AND BOTH ED BATTERIES ARE GO.
TWENTY-FIVE.
RATE SCALE, TWENTY-FIVE?
ATT TRANSLATION, FOUR JETS.
BALANCE COUPLE ON.
BALANCE COUPLE ON.
EAGLE, HOUSTON.
YOU'RE LOOKING GOOD TO US.
ROGER.
MARK.
TIG* MINUS 2.
* TIME OF IGNITION.
OK.
MASTER ARM ON!
NINE, EIGHT, SEVEN, SIX, FIVE, ABORT STAGE...
...ENGINE ARM ASCENT, PROCEED.

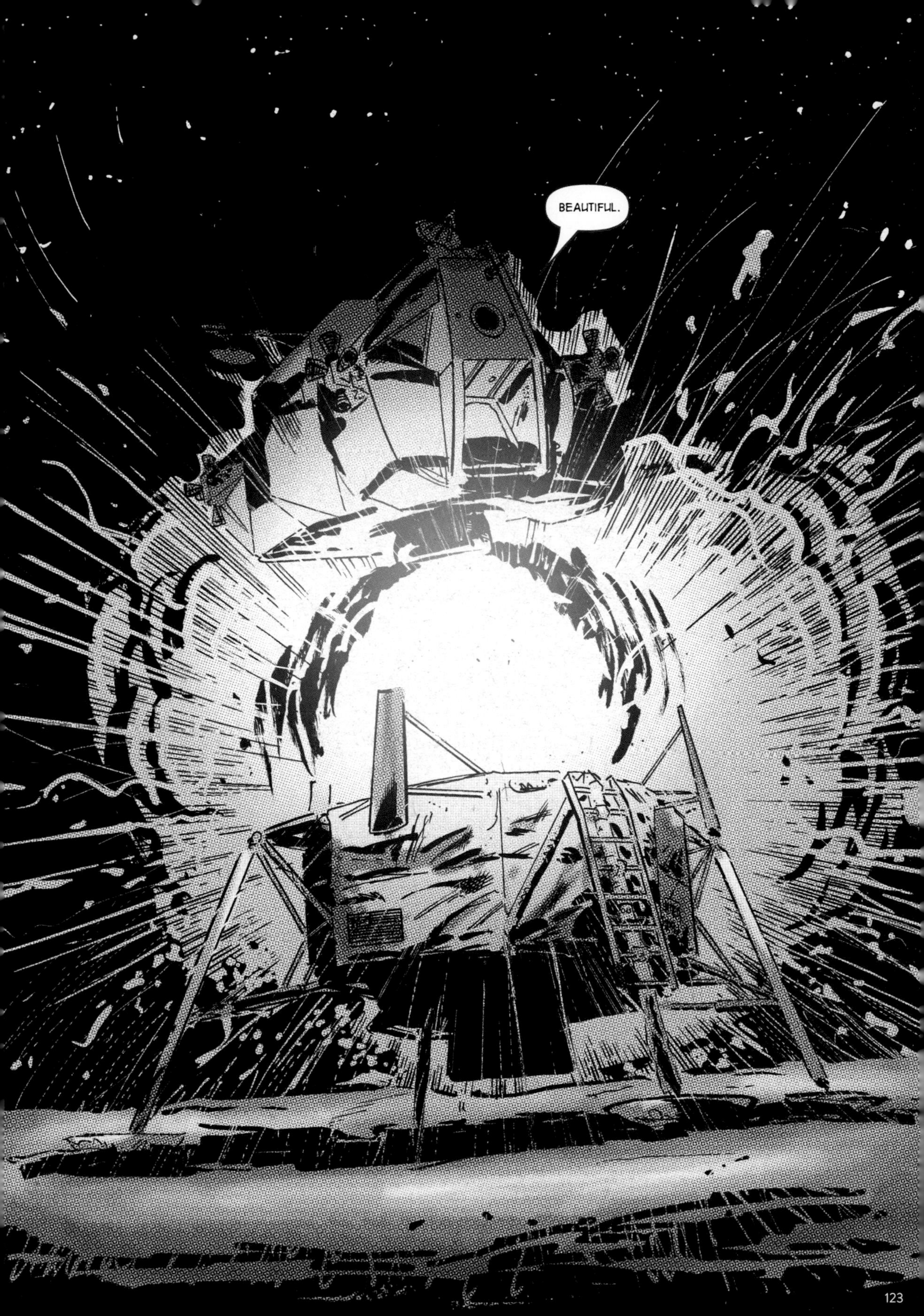
BEAUTIFUL.

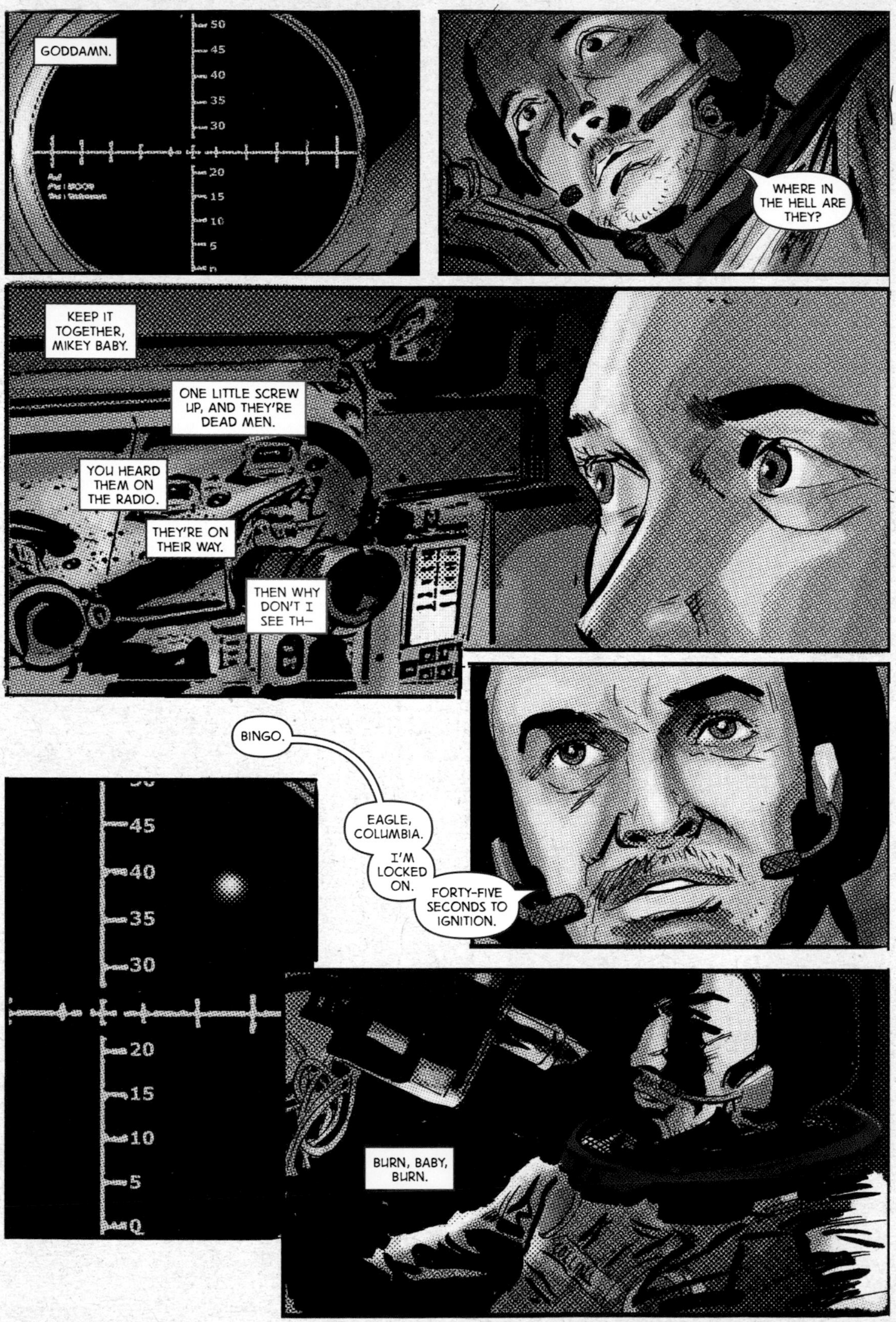
GODDAMN.
WHERE IN THE HELL ARE THEY?
KEEP IT TOGETHER, MIKEY BABY.
ONE LITTLE SCREW UP, AND THEY'RE DEAD MEN.
YOU HEARD THEM ON THE RADIO.
THEY'RE ON THEIR WAY.
THEN WHY DON'T I SEE TH–
BINGO.
EAGLE, COLUMBIA.
I'M LOCKED ON.
FORTY-FIVE SECONDS TO IGNITION.
BURN, BABY, BURN.

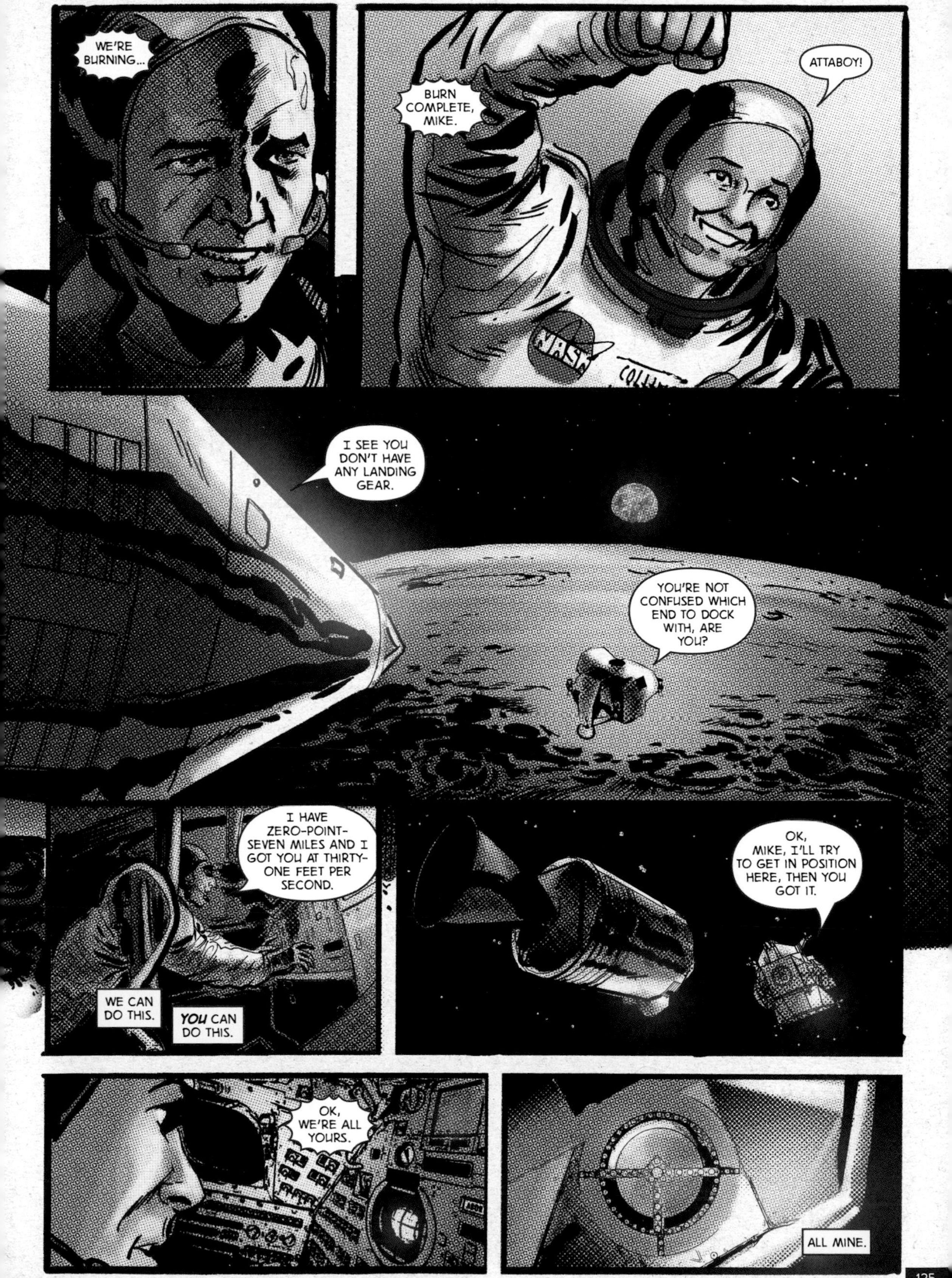
WE'RE BURNING...
BURN COMPLETE, MIKE.
ATTABOY!
I SEE YOU DON'T HAVE ANY LANDING GEAR.
YOU'RE NOT CONFUSED WHICH END TO DOCK WITH, ARE YOU?
I HAVE ZERO-POINT-SEVEN MILES AND I GOT YOU AT THIRTY-ONE FEET PER SECOND.
WE CAN DO THIS.
YOU CAN DO THIS.
OK, MIKE, I'LL TRY TO GET IN POSITION HERE, THEN YOU GOT IT.
OK, WE'RE ALL YOURS.
ALL MINE.

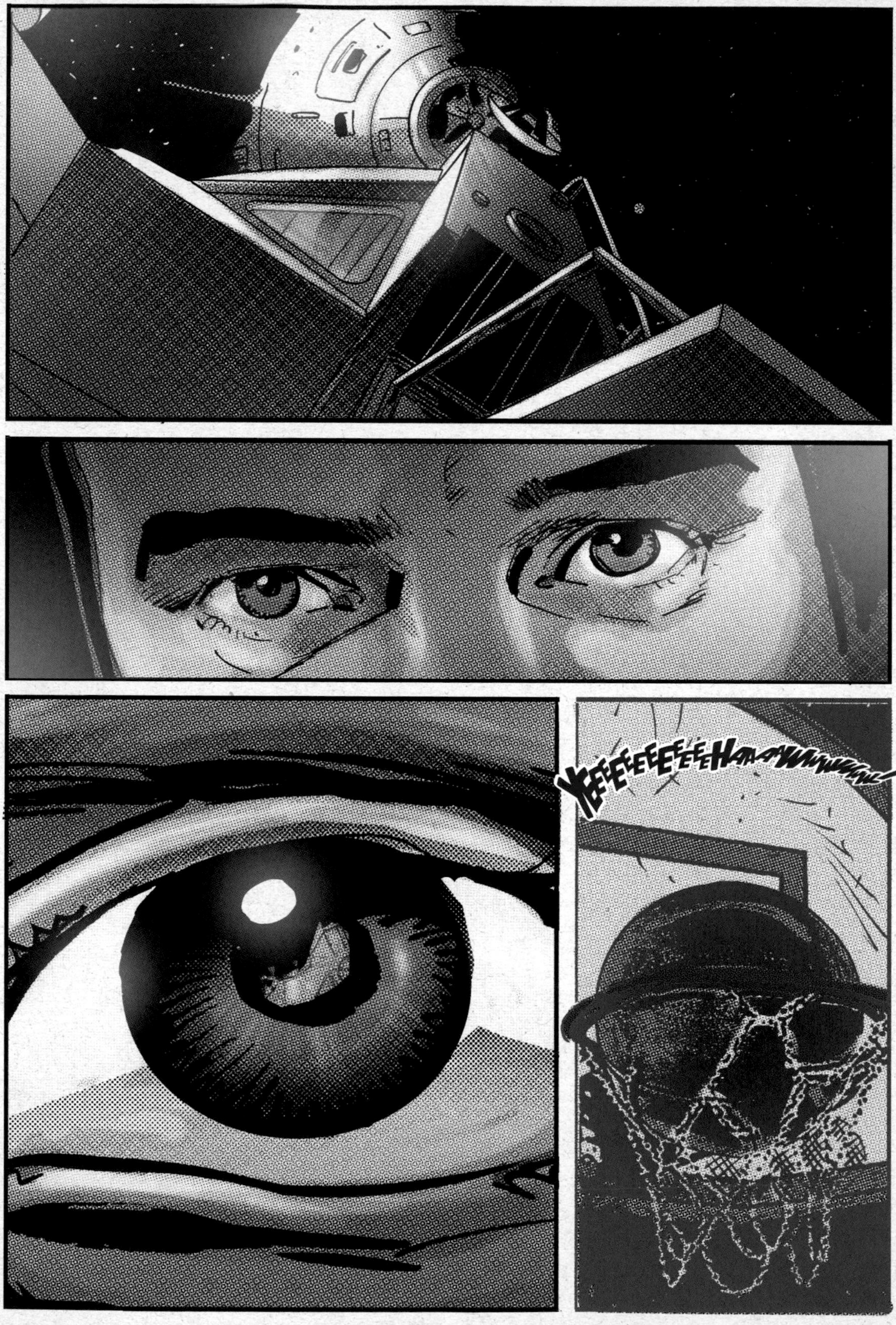
YEEEEEEEEEEHAAAWWWWWWW!

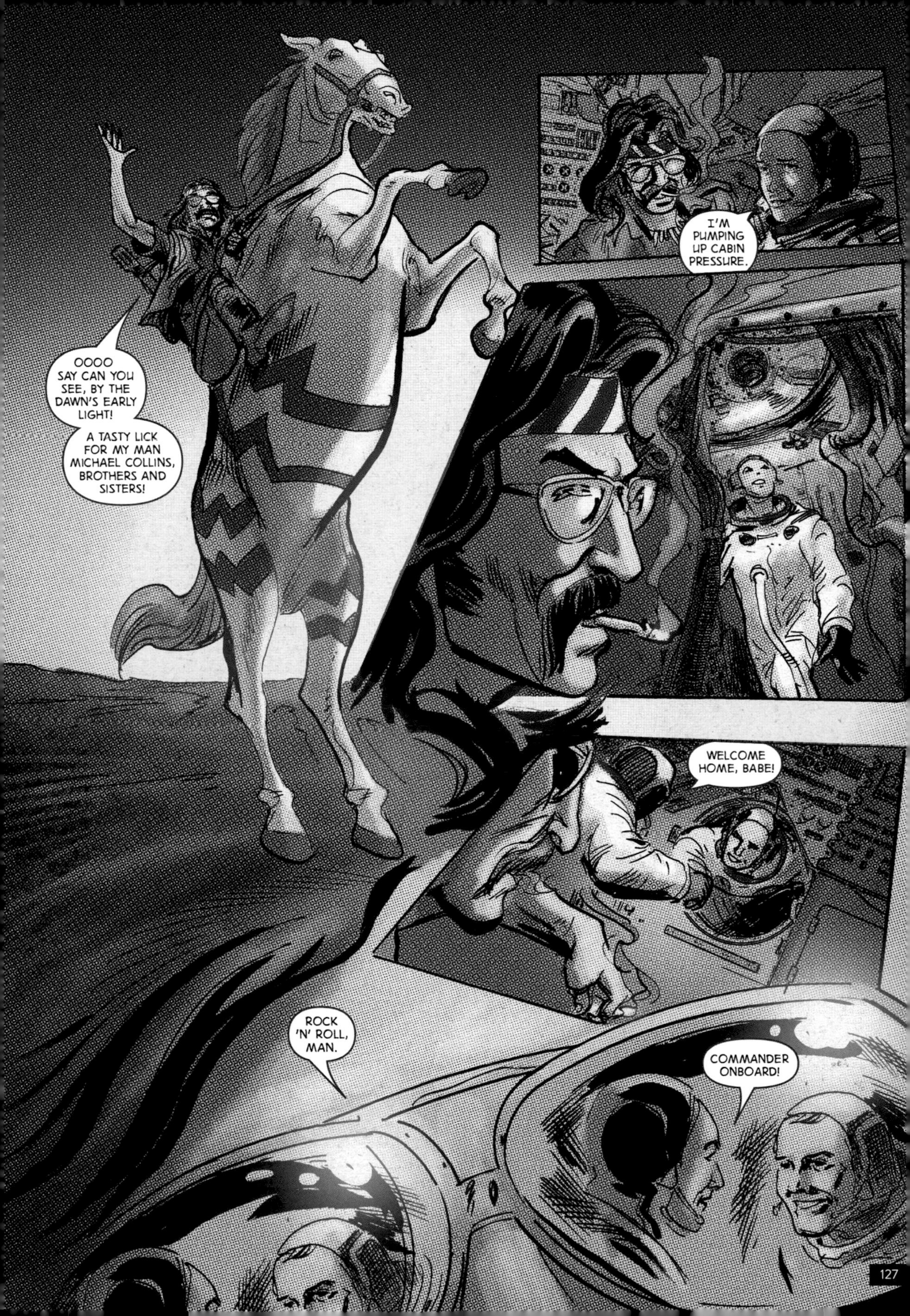
OOOO SAY CAN YOU SEE, BY THE DAWN'S EARLY LIGHT!
A TASTY LICK FOR MY MAN MICHAEL COLLINS, BROTHERS AND SISTERS!
I'M PUMPING UP CABIN PRESSURE.
WELCOME HOME, BABE!
ROCK 'N' ROLL, MAN.
COMMANDER ONBOARD!

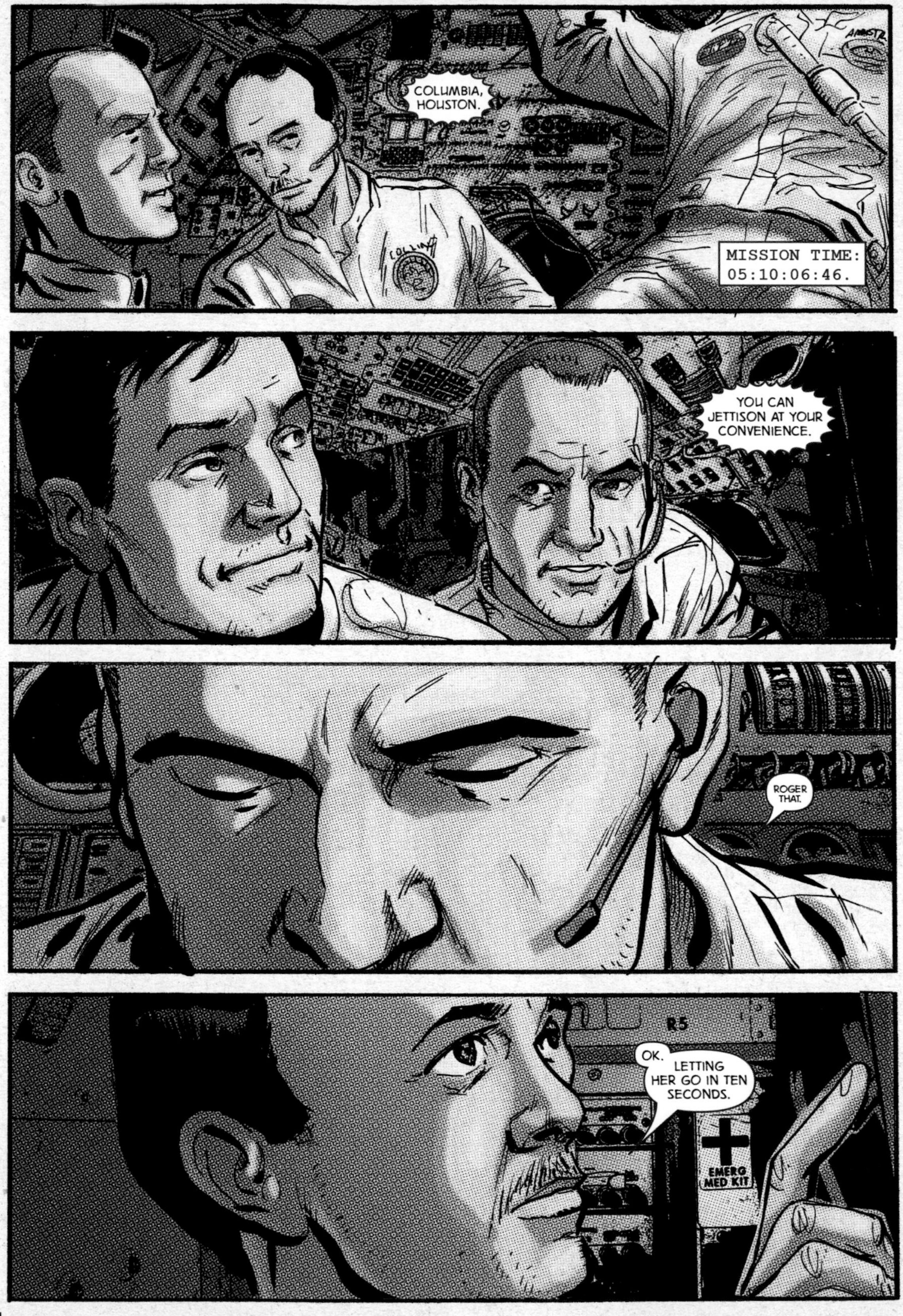
COLUMBIA, HOUSTON.
MISSION TIME: 05:10:06:46.
YOU CAN JETTISON AT YOUR CONVENIENCE.
ROGER THAT.
OK.
LETTING HER GO IN TEN SECONDS.
R5
EMERG MED KIT

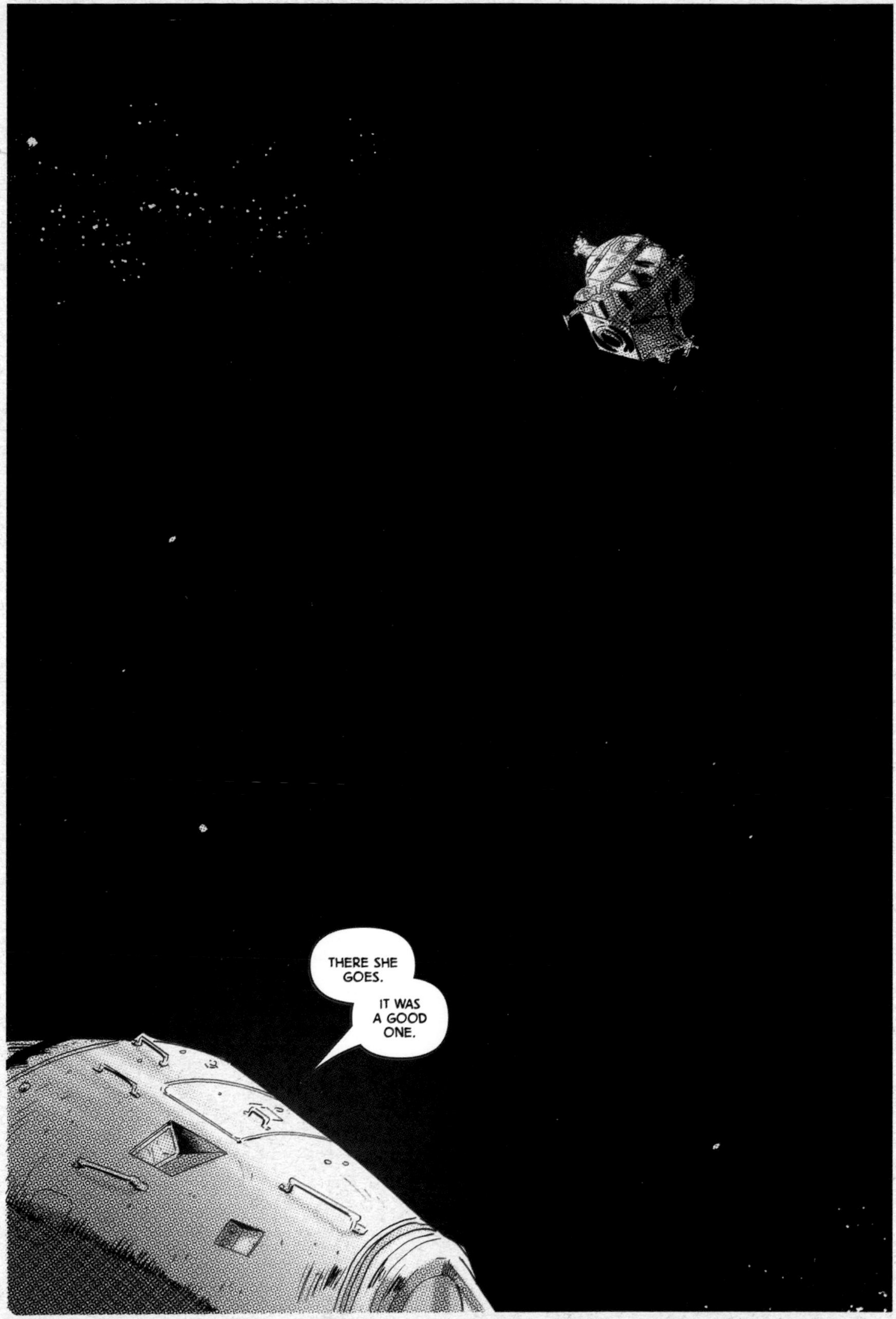
THERE SHE GOES.
IT WAS A GOOD ONE.

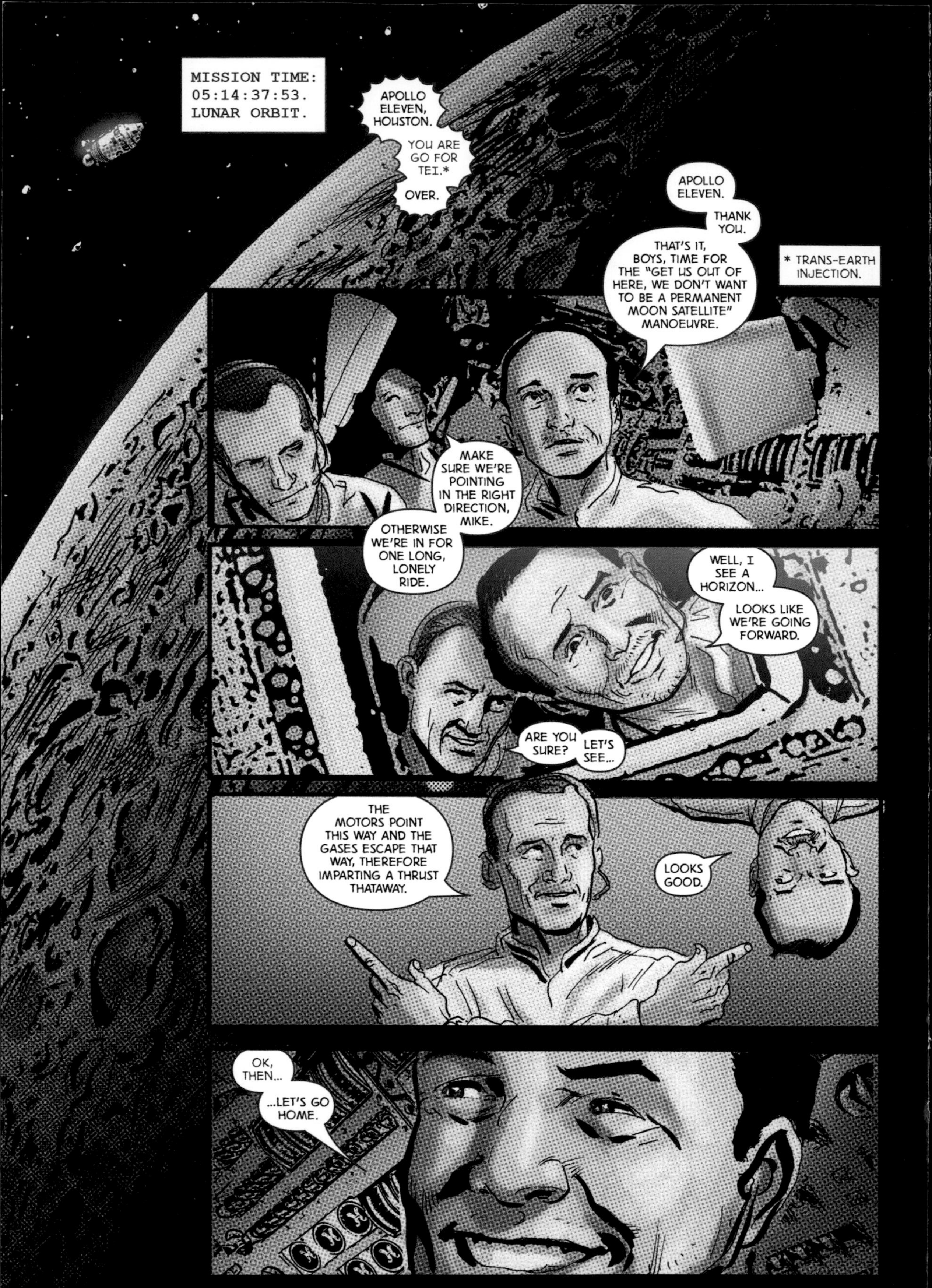
MISSION TIME: 05:14:37:53. LUNAR ORBIT.
APOLLO ELEVEN, HOUSTON.
YOU ARE GO FOR TEI.*
OVER.
APOLLO ELEVEN.
THANK YOU.
THAT'S IT, BOYS, TIME FOR THE "GET US OUT OF HERE, WE DON'T WANT TO BE A PERMANENT MOON SATELLITE" MANOEUVRE.
* TRANS-EARTH INJECTION.
MAKE SURE WE'RE POINTING IN THE RIGHT DIRECTION, MIKE.
OTHERWISE WE'RE IN FOR ONE LONG, LONELY RIDE.
WELL, I SEE A HORIZON...
LOOKS LIKE WE'RE GOING FORWARD.
ARE YOU SURE?
LET'S SEE...
THE MOTORS POINT THIS WAY AND THE GASES ESCAPE THAT WAY, THEREFORE IMPARTING A THRUST THATAWAY.
LOOKS GOOD.
OK, THEN...
...LET'S GO HOME.

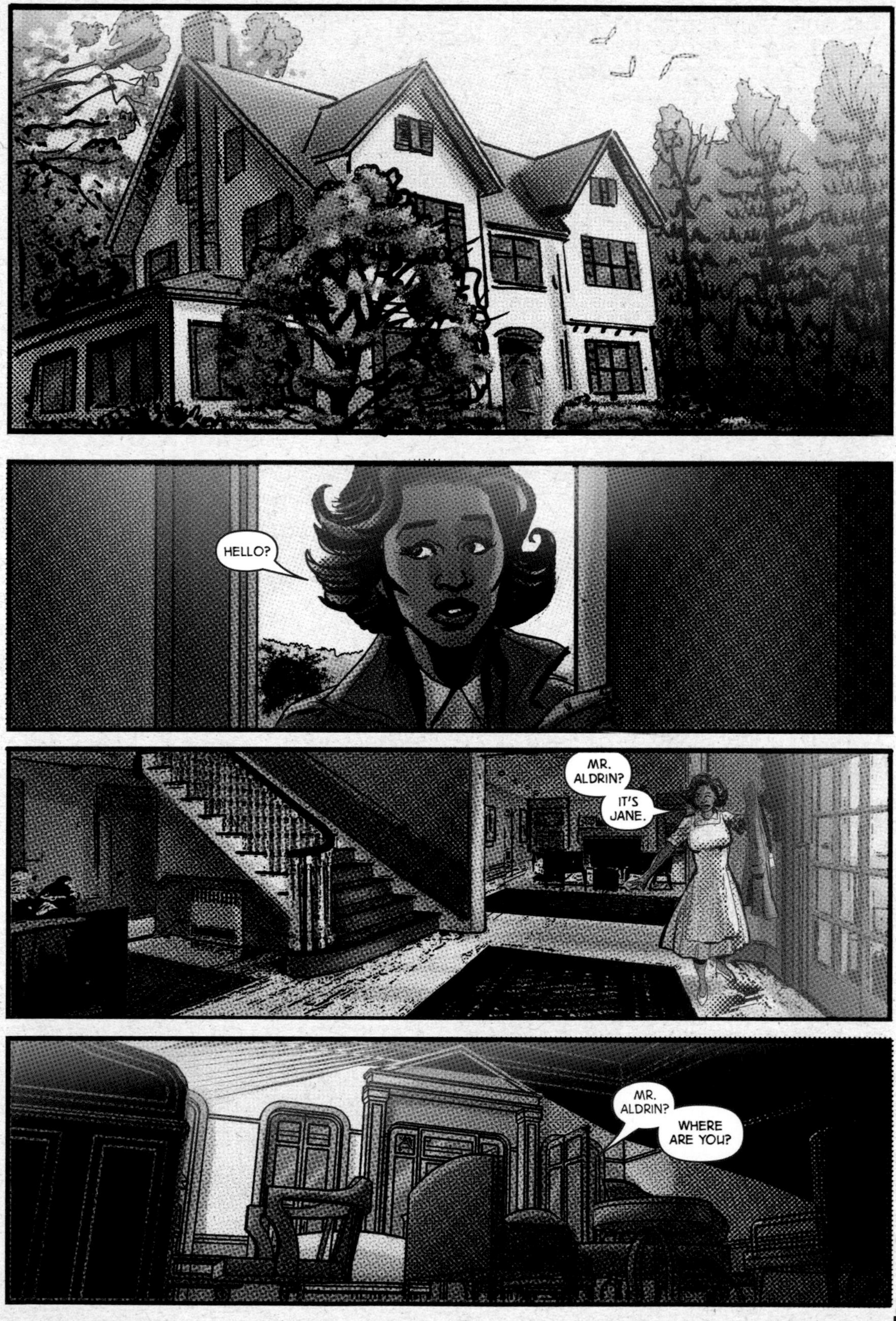
HELLO?
MR. ALDRIN?
IT'S JANE.
MR. ALDRIN?
WHERE ARE YOU?

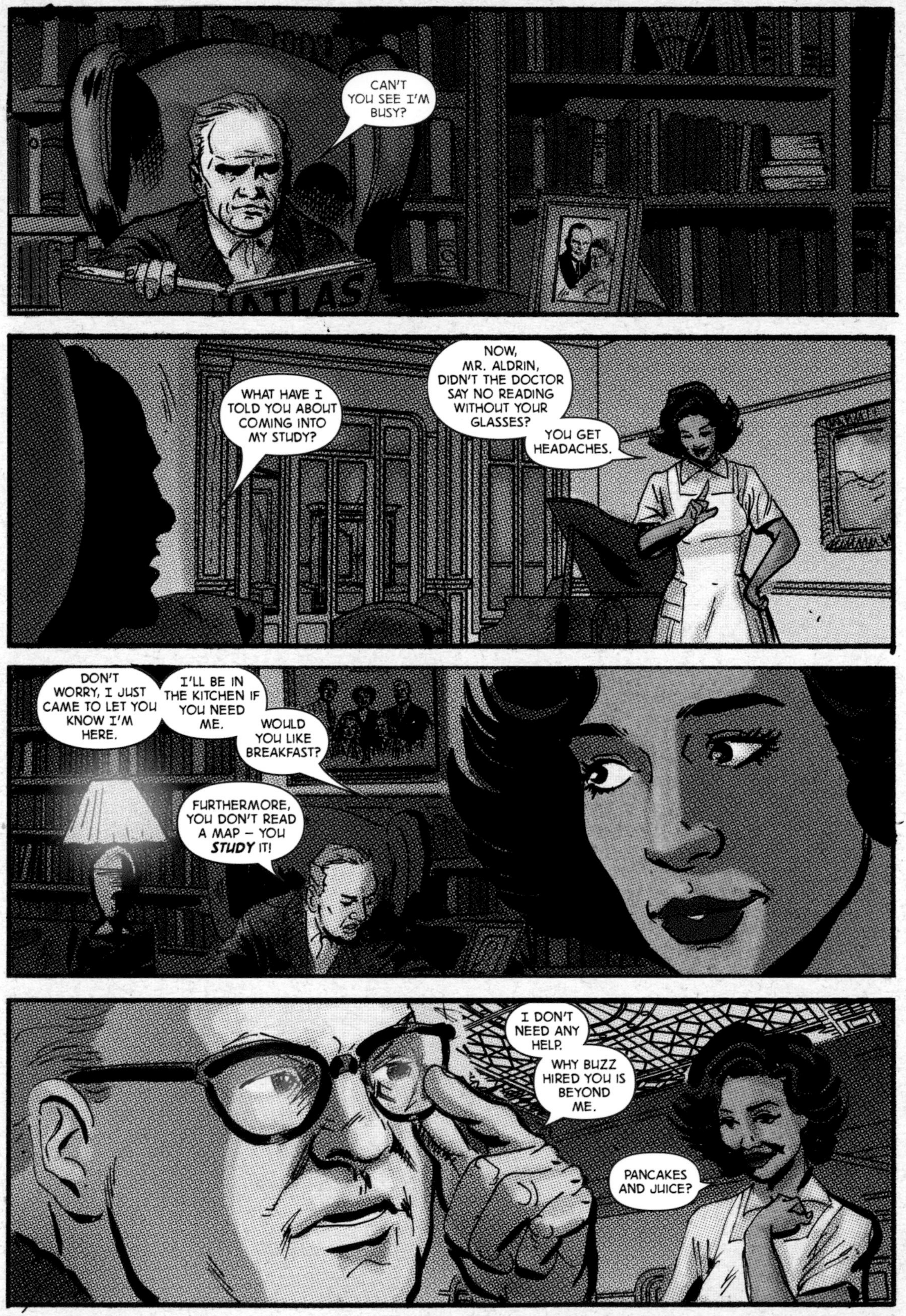
CAN'T YOU SEE I'M BUSY?
ATLAS
WHAT HAVE I TOLD YOU ABOUT COMING INTO MY STUDY?
NOW, MR. ALDRIN, DIDN'T THE DOCTOR SAY NO READING WITHOUT YOUR GLASSES?
YOU GET HEADACHES.
DON'T WORRY, I JUST CAME TO LET YOU KNOW I'M HERE.
I'LL BE IN THE KITCHEN IF YOU NEED ME.
WOULD YOU LIKE BREAKFAST?
FURTHERMORE, YOU DON'T READ A MAP – YOU *STUDY* IT!
I DON'T NEED ANY HELP.
WHY BUZZ HIRED YOU IS BEYOND ME.
PANCAKES AND JUICE?

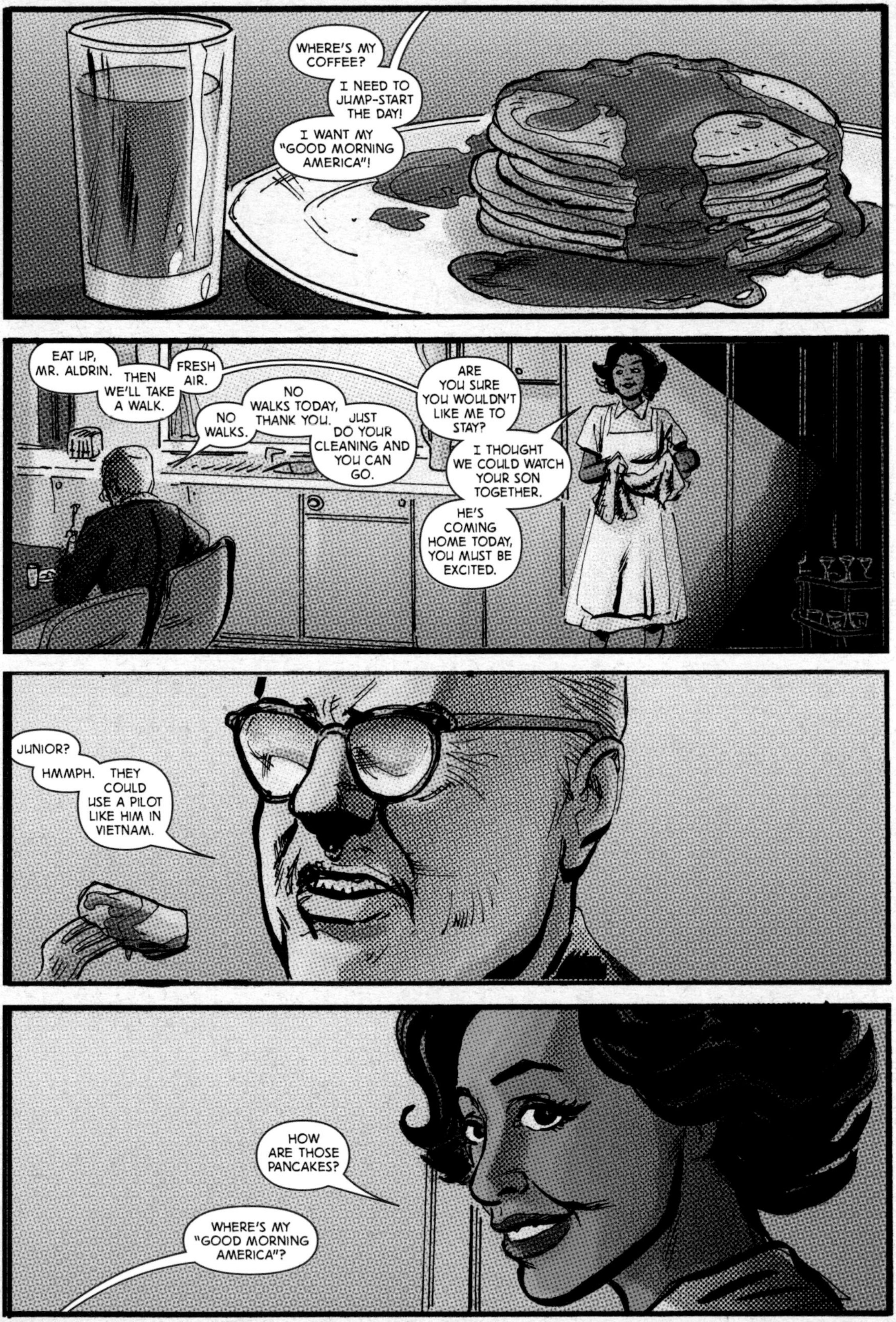
WHERE'S MY COFFEE?
I NEED TO JUMP-START THE DAY!
I WANT MY "GOOD MORNING AMERICA"!
EAT UP, MR. ALDRIN.
THEN WE'LL TAKE A WALK.
FRESH AIR.
NO WALKS.
NO WALKS TODAY, THANK YOU.
JUST DO YOUR CLEANING AND YOU CAN GO.
ARE YOU SURE YOU WOULDN'T LIKE ME TO STAY?
I THOUGHT WE COULD WATCH YOUR SON TOGETHER.
HE'S COMING HOME TODAY, YOU MUST BE EXCITED.
JUNIOR?
HMMPH. THEY COULD USE A PILOT LIKE HIM IN VIETNAM.
HOW ARE THOSE PANCAKES?
WHERE'S MY "GOOD MORNING AMERICA"?

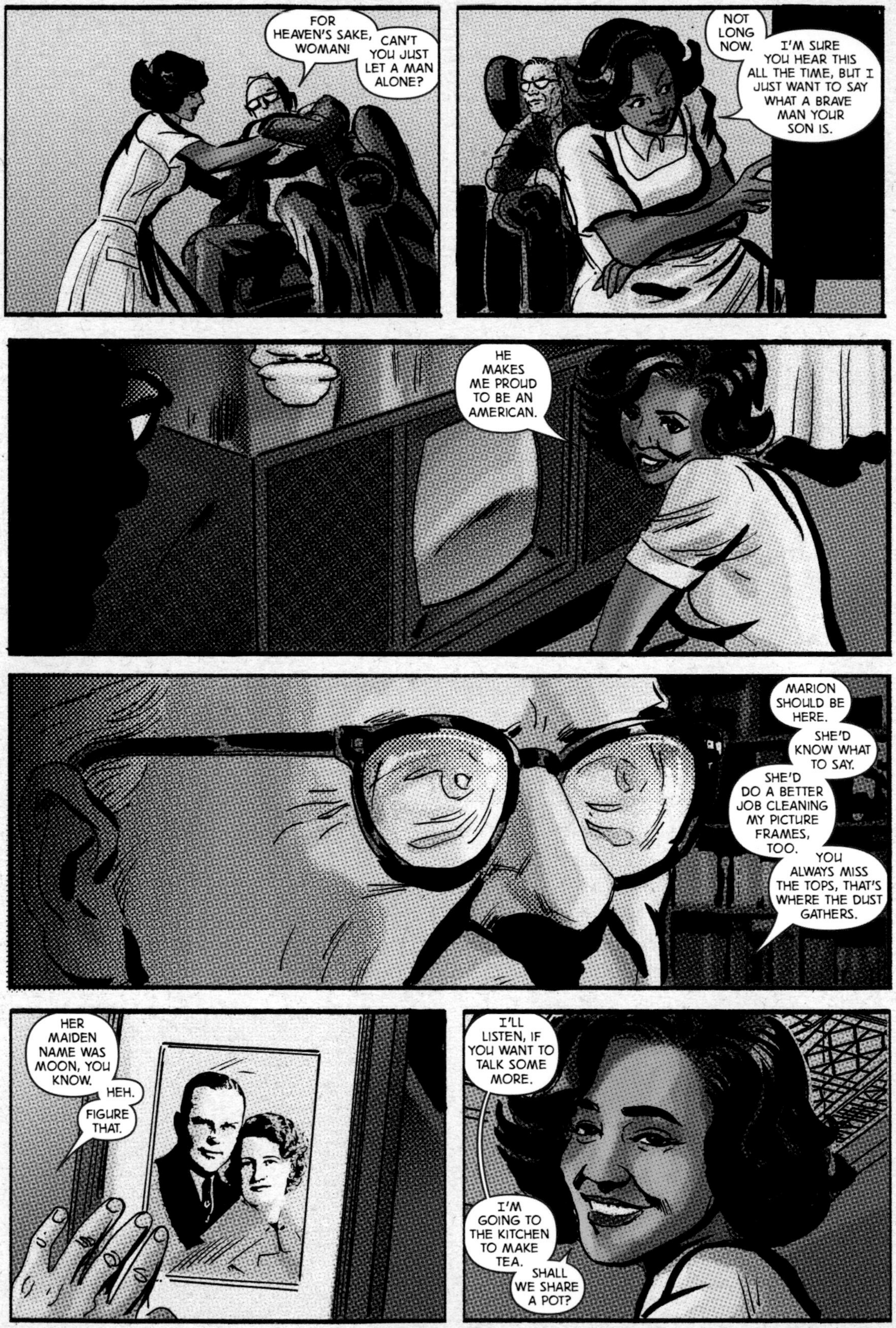
FOR HEAVEN'S SAKE, WOMAN!
CAN'T YOU JUST LET A MAN ALONE?
NOT LONG NOW.
I'M SURE YOU HEAR THIS ALL THE TIME, BUT I JUST WANT TO SAY WHAT A BRAVE MAN YOUR SON IS.
HE MAKES ME PROUD TO BE AN AMERICAN.
MARION SHOULD BE HERE.
SHE'D KNOW WHAT TO SAY.
SHE'D DO A BETTER JOB CLEANING MY PICTURE FRAMES, TOO.
YOU ALWAYS MISS THE TOPS, THAT'S WHERE THE DUST GATHERS.
HER MAIDEN NAME WAS MOON, YOU KNOW.
HEH.
FIGURE THAT.
I'LL LISTEN, IF YOU WANT TO TALK SOME MORE.
I'M GOING TO THE KITCHEN TO MAKE TEA.
SHALL WE SHARE A POT?

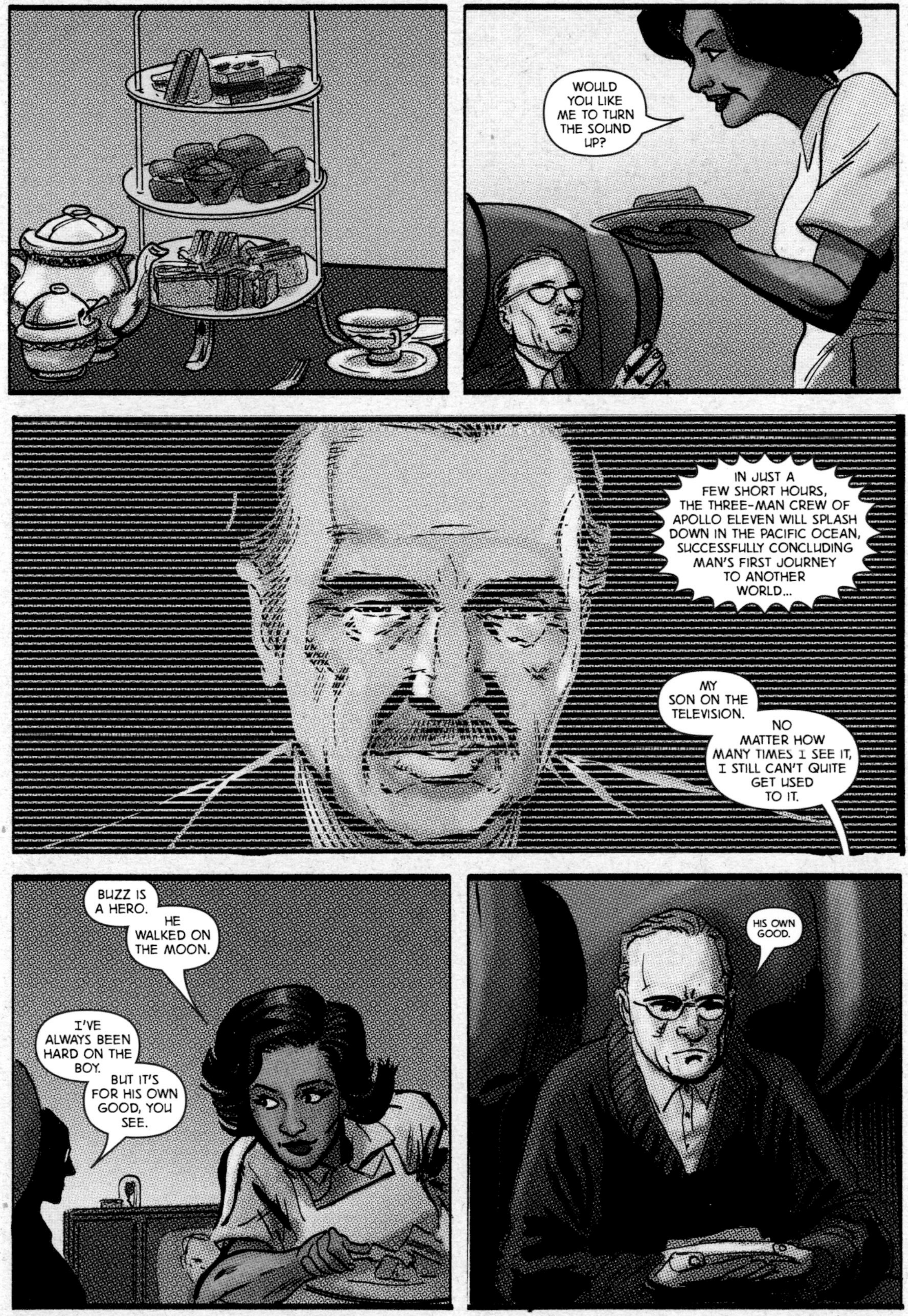
WOULD YOU LIKE ME TO TURN THE SOUND UP?
IN JUST A FEW SHORT HOURS, THE THREE-MAN CREW OF APOLLO ELEVEN WILL SPLASH DOWN IN THE PACIFIC OCEAN, SUCCESSFULLY CONCLUDING MAN'S FIRST JOURNEY TO ANOTHER WORLD...
MY SON ON THE TELEVISION.
NO MATTER HOW MANY TIMES I SEE IT, I STILL CAN'T QUITE GET USED TO IT.
BUZZ IS A HERO.
HE WALKED ON THE MOON.
I'VE ALWAYS BEEN HARD ON THE BOY.
BUT IT'S FOR HIS OWN GOOD, YOU SEE.
HIS OWN GOOD.

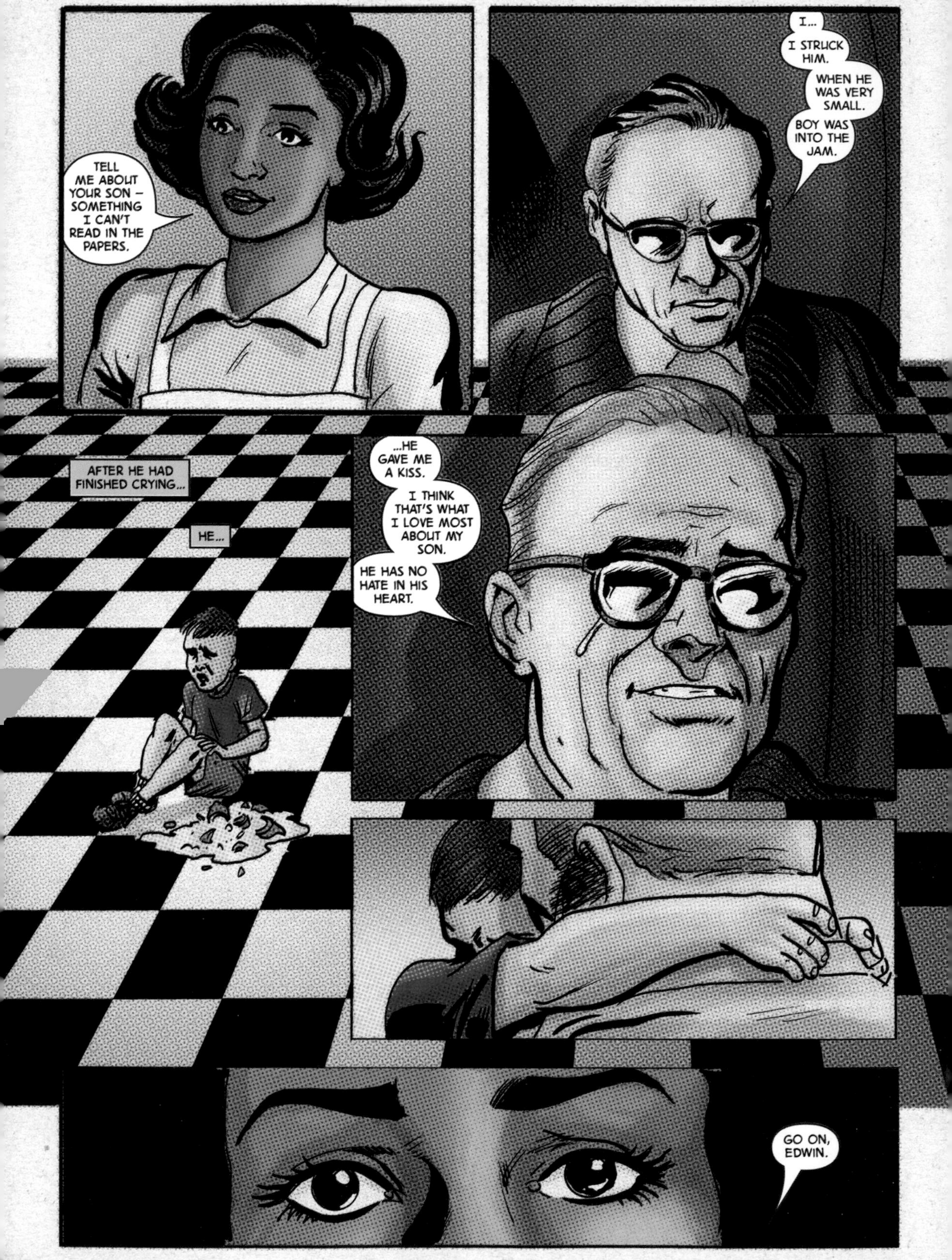
TELL ME ABOUT YOUR SON – SOMETHING I CAN'T READ IN THE PAPERS.
I...
I STRUCK HIM.
WHEN HE WAS VERY SMALL.
BOY WAS INTO THE JAM.
AFTER HE HAD FINISHED CRYING...
HE...
...HE GAVE ME A KISS.
I THINK THAT'S WHAT I LOVE MOST ABOUT MY SON.
HE HAS NO HATE IN HIS HEART.
GO ON, EDWIN.

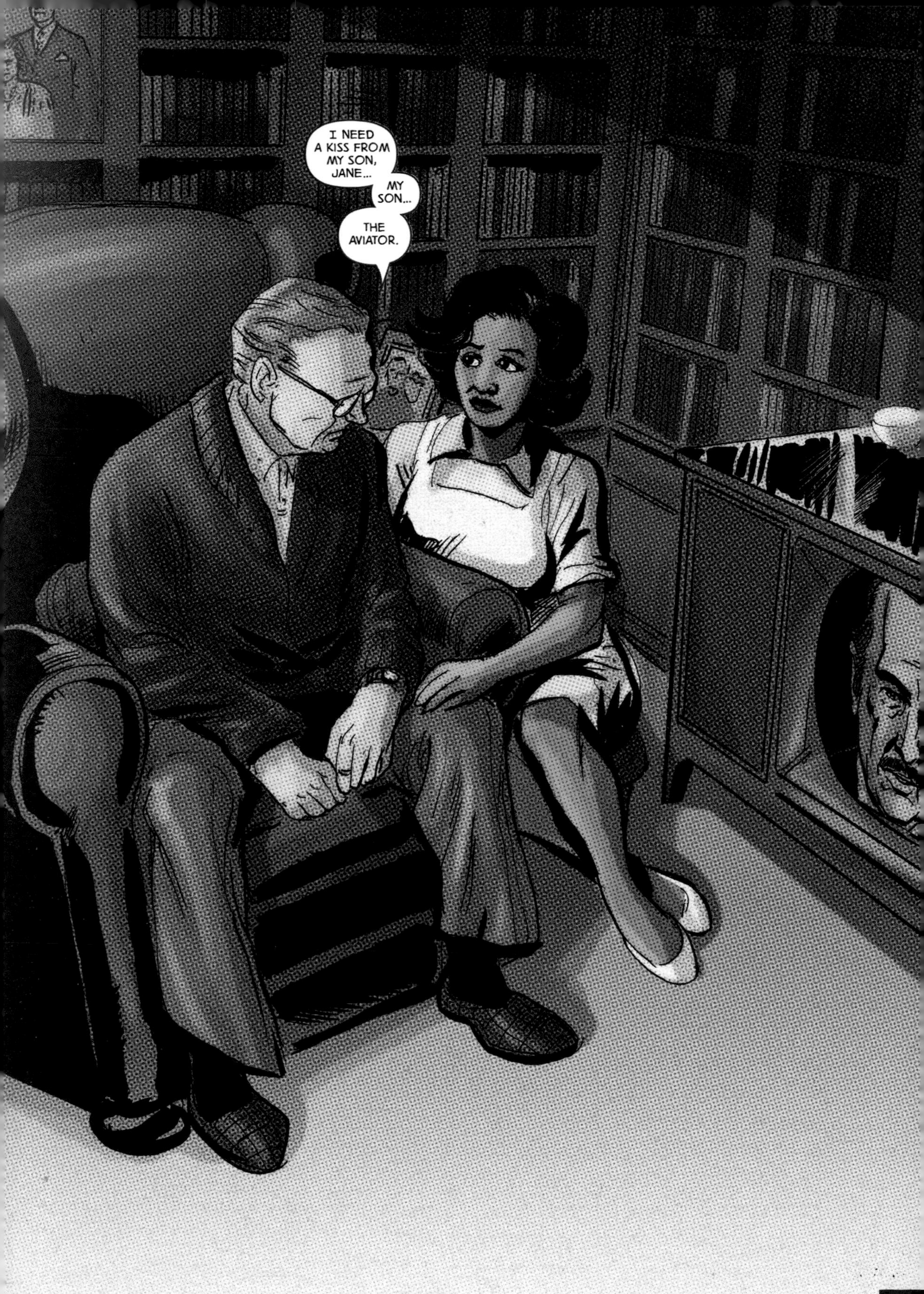
I NEED A KISS FROM MY SON, JANE...
MY SON...
THE AVIATOR.

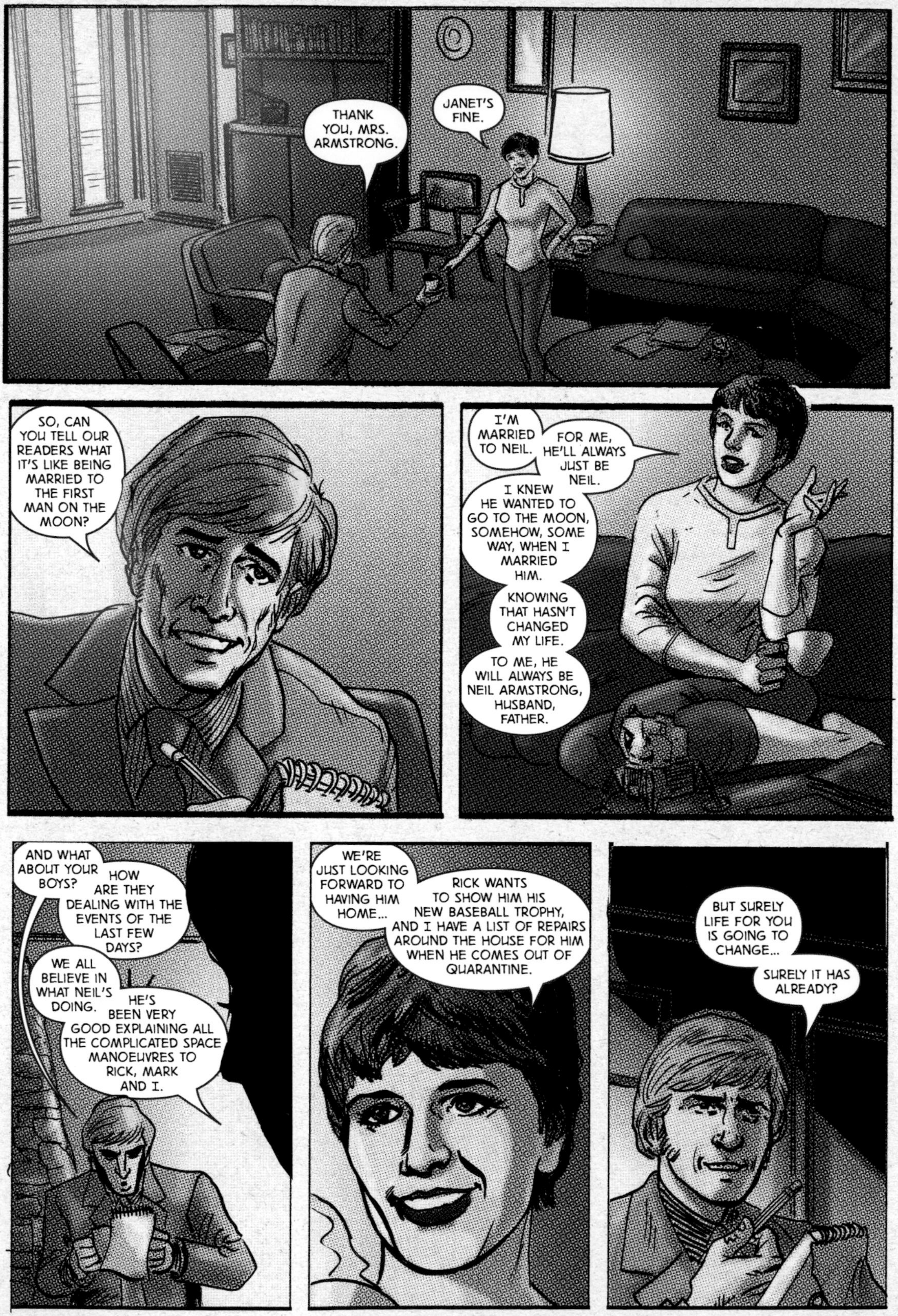
THANK YOU, MRS. ARMSTRONG.
JANET'S FINE.
SO, CAN YOU TELL OUR READERS WHAT IT'S LIKE BEING MARRIED TO THE FIRST MAN ON THE MOON?
I'M MARRIED TO NEIL.
FOR ME, HE'LL ALWAYS JUST BE NEIL.
I KNEW HE WANTED TO GO TO THE MOON, SOMEHOW, SOME WAY, WHEN I MARRIED HIM.
KNOWING THAT HASN'T CHANGED MY LIFE.
TO ME, HE WILL ALWAYS BE NEIL ARMSTRONG, HUSBAND, FATHER.
AND WHAT ABOUT YOUR BOYS?
HOW ARE THEY DEALING WITH THE EVENTS OF THE LAST FEW DAYS?
WE ALL BELIEVE IN WHAT NEIL'S DOING.
HE'S BEEN VERY GOOD EXPLAINING ALL THE COMPLICATED SPACE MANOEUVRES TO RICK, MARK AND I.
WE'RE JUST LOOKING FORWARD TO HAVING HIM HOME...
RICK WANTS TO SHOW HIM HIS NEW BASEBALL TROPHY, AND I HAVE A LIST OF REPAIRS AROUND THE HOUSE FOR HIM WHEN HE COMES OUT OF QUARANTINE.
BUT SURELY LIFE FOR YOU IS GOING TO CHANGE...
SURELY IT HAS ALREADY?

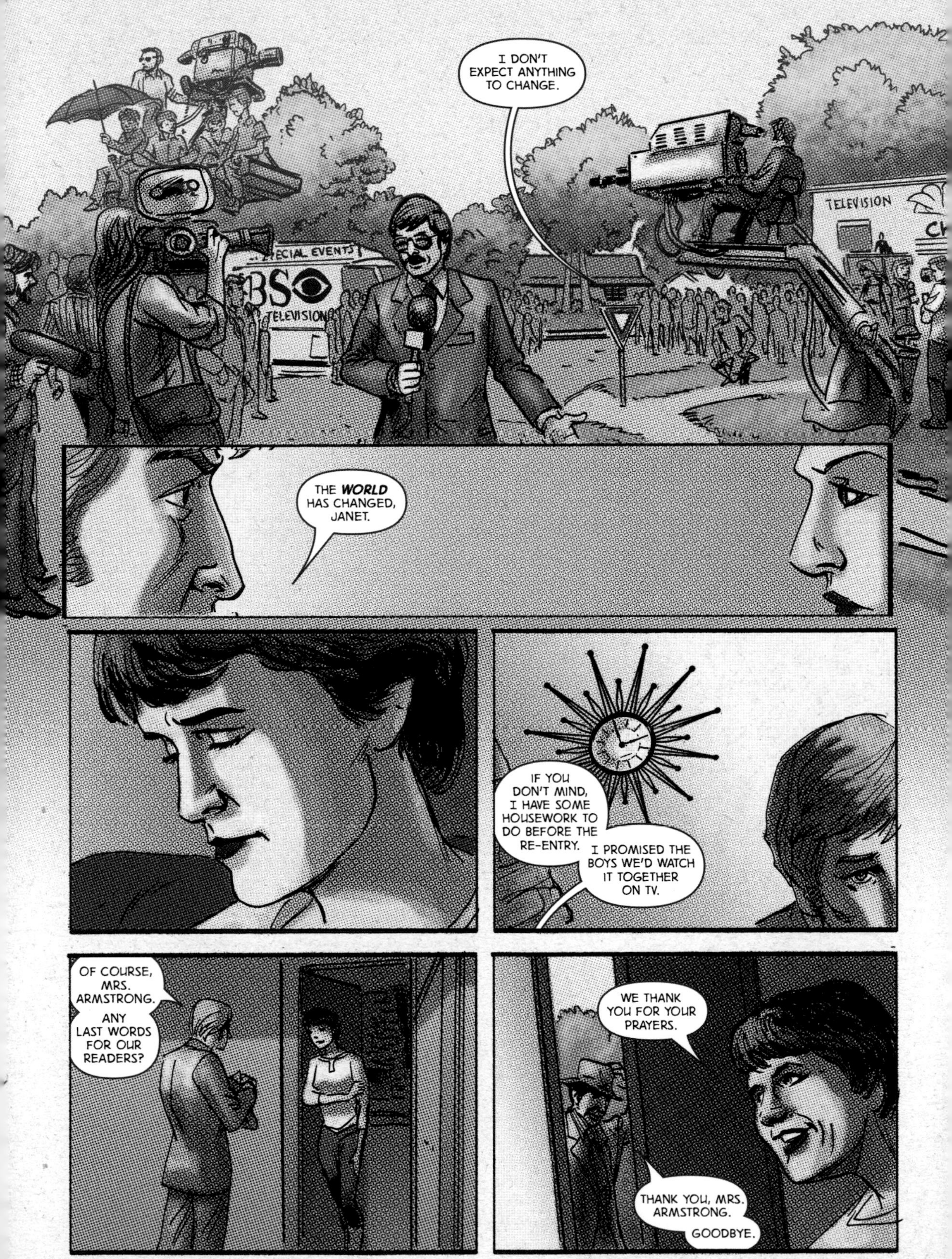

I DON'T EXPECT ANYTHING TO CHANGE.
SPECIAL EVENTS
CBS
TELEVISION
TELEVISION
THE *WORLD* HAS CHANGED, JANET.
IF YOU DON'T MIND, I HAVE SOME HOUSEWORK TO DO BEFORE THE RE-ENTRY.
I PROMISED THE BOYS WE'D WATCH IT TOGETHER ON TV.
OF COURSE, MRS. ARMSTRONG.
ANY LAST WORDS FOR OUR READERS?
WE THANK YOU FOR YOUR PRAYERS.
THANK YOU, MRS. ARMSTRONG.
GOODBYE.

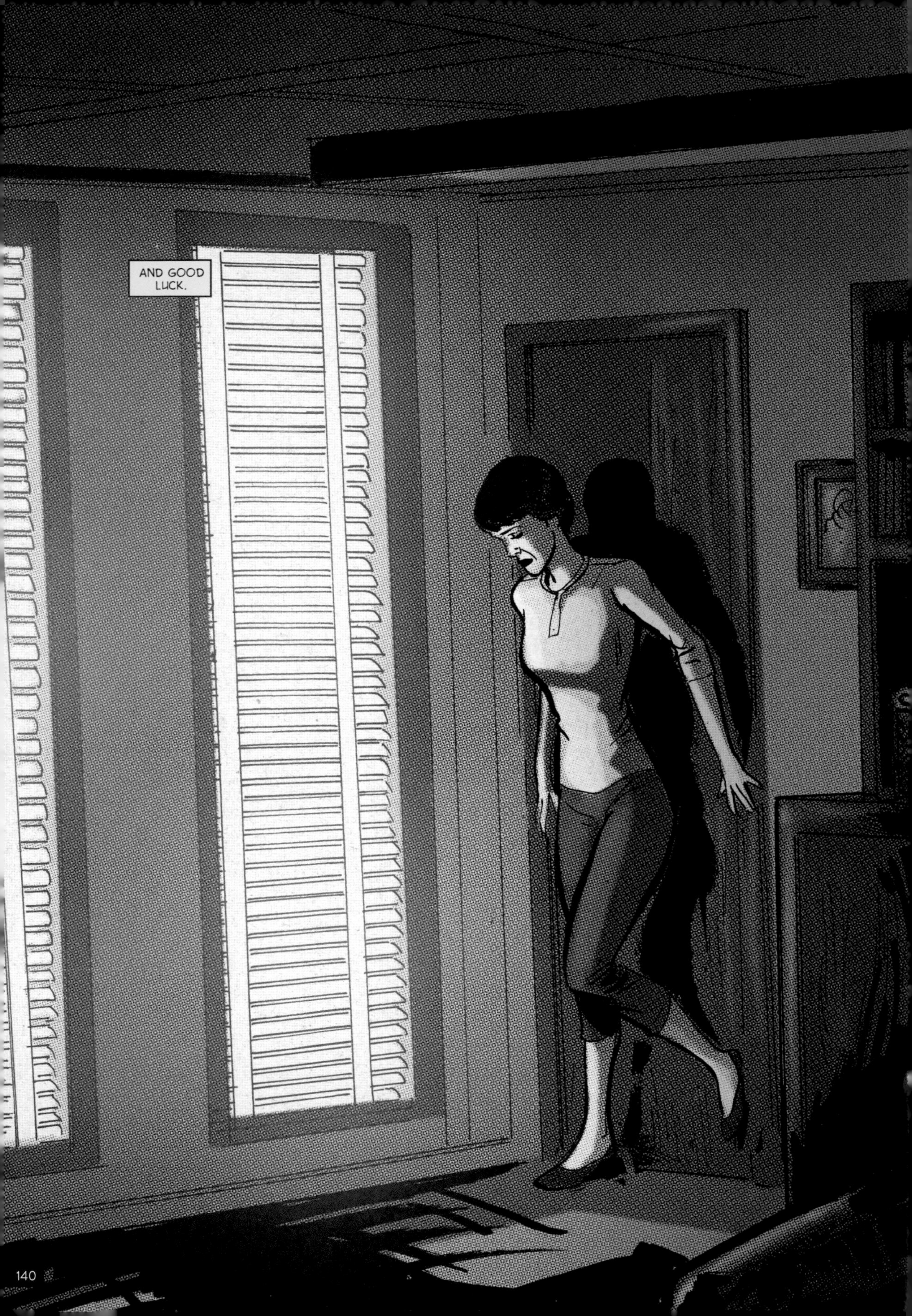
AND GOOD LUCK.

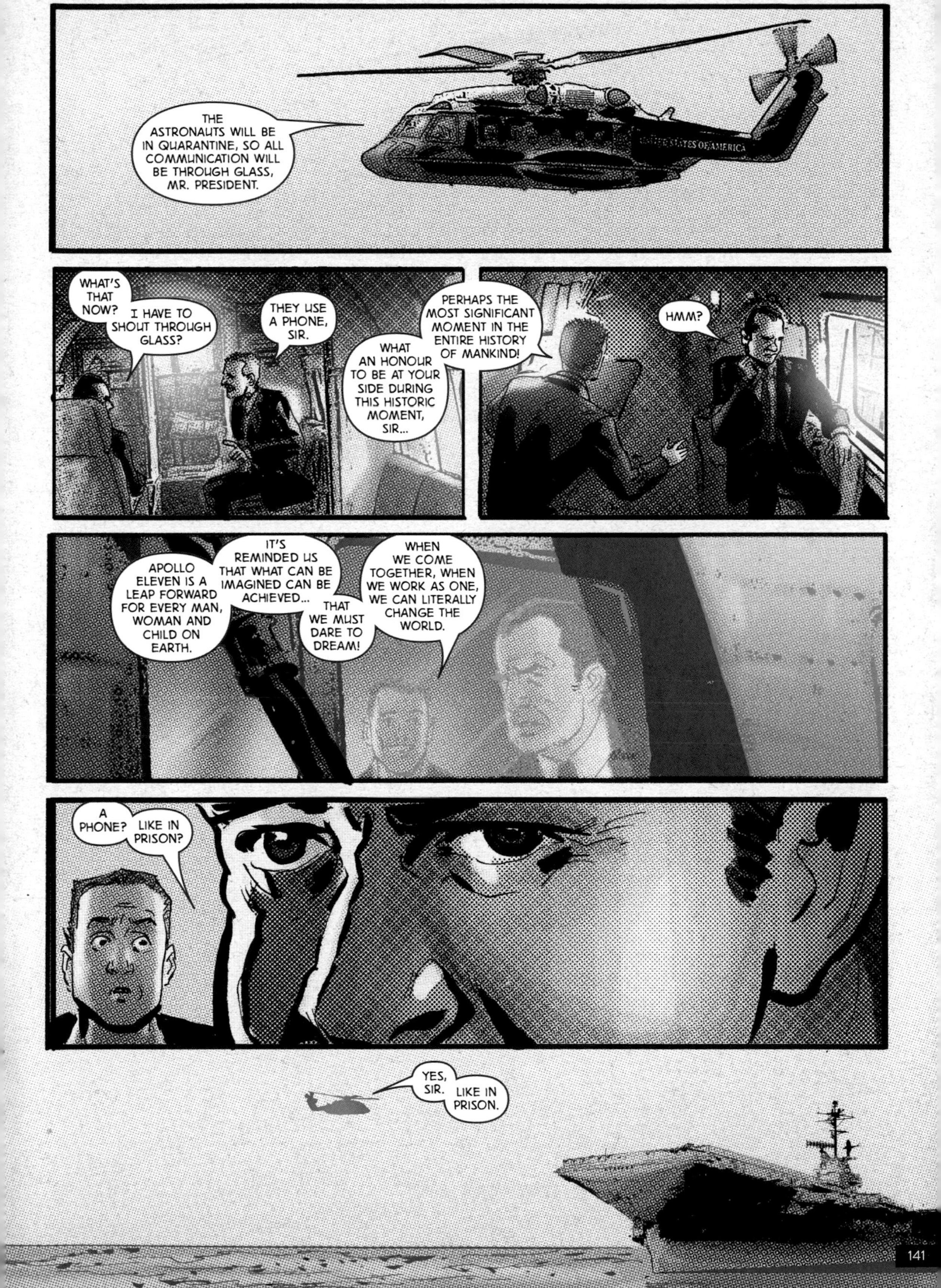
THE ASTRONAUTS WILL BE IN QUARANTINE, SO ALL COMMUNICATION WILL BE THROUGH GLASS, MR. PRESIDENT.
UNITED STATES OF AMERICA
WHAT'S THAT NOW?
I HAVE TO SHOUT THROUGH GLASS?
THEY USE A PHONE, SIR.
WHAT AN HONOUR TO BE AT YOUR SIDE DURING THIS HISTORIC MOMENT, SIR...
PERHAPS THE MOST SIGNIFICANT MOMENT IN THE ENTIRE HISTORY OF MANKIND!
HMM?
APOLLO ELEVEN IS A LEAP FORWARD FOR EVERY MAN, WOMAN AND CHILD ON EARTH.
IT'S REMINDED US THAT WHAT CAN BE IMAGINED CAN BE ACHIEVED...
THAT WE MUST DARE TO DREAM!
WHEN WE COME TOGETHER, WHEN WE WORK AS ONE, WE CAN LITERALLY CHANGE THE WORLD.
A PHONE?
LIKE IN PRISON?
YES, SIR.
LIKE IN PRISON.

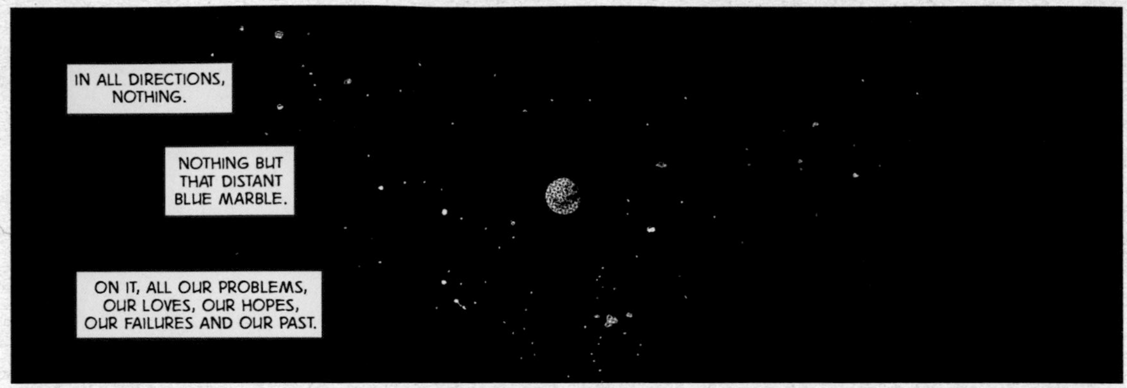
IN ALL DIRECTIONS, NOTHING.
NOTHING BUT THAT DISTANT BLUE MARBLE.
ON IT, ALL OUR PROBLEMS, OUR LOVES, OUR HOPES, OUR FAILURES AND OUR PAST.

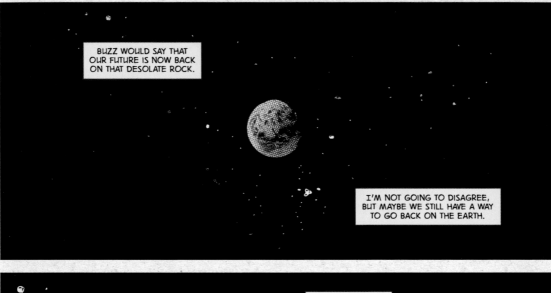

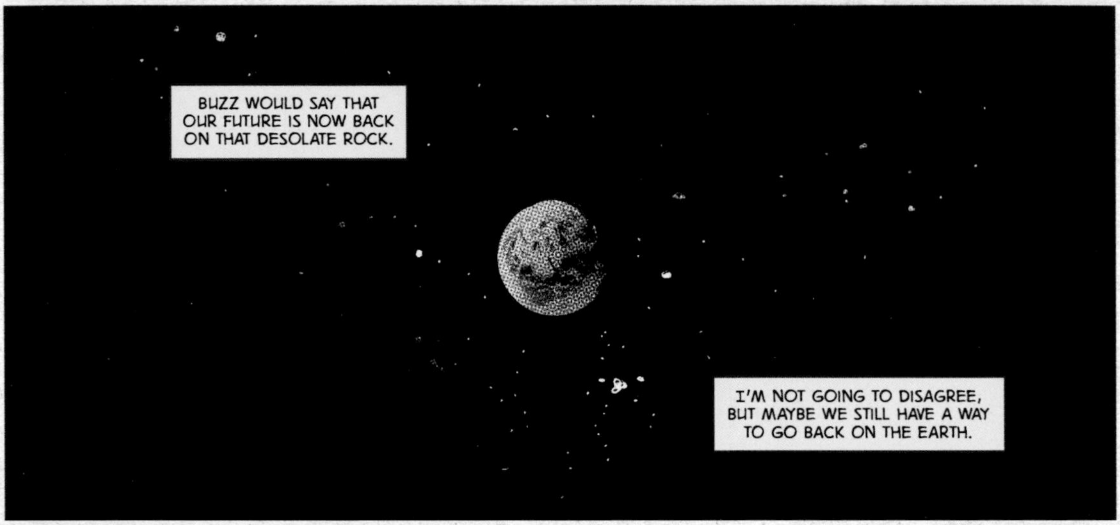
BUZZ WOULD SAY THAT OUR FUTURE IS NOW BACK ON THAT DESOLATE ROCK.
I'M NOT GOING TO DISAGREE, BUT MAYBE WE STILL HAVE A WAY TO GO BACK ON THE EARTH.

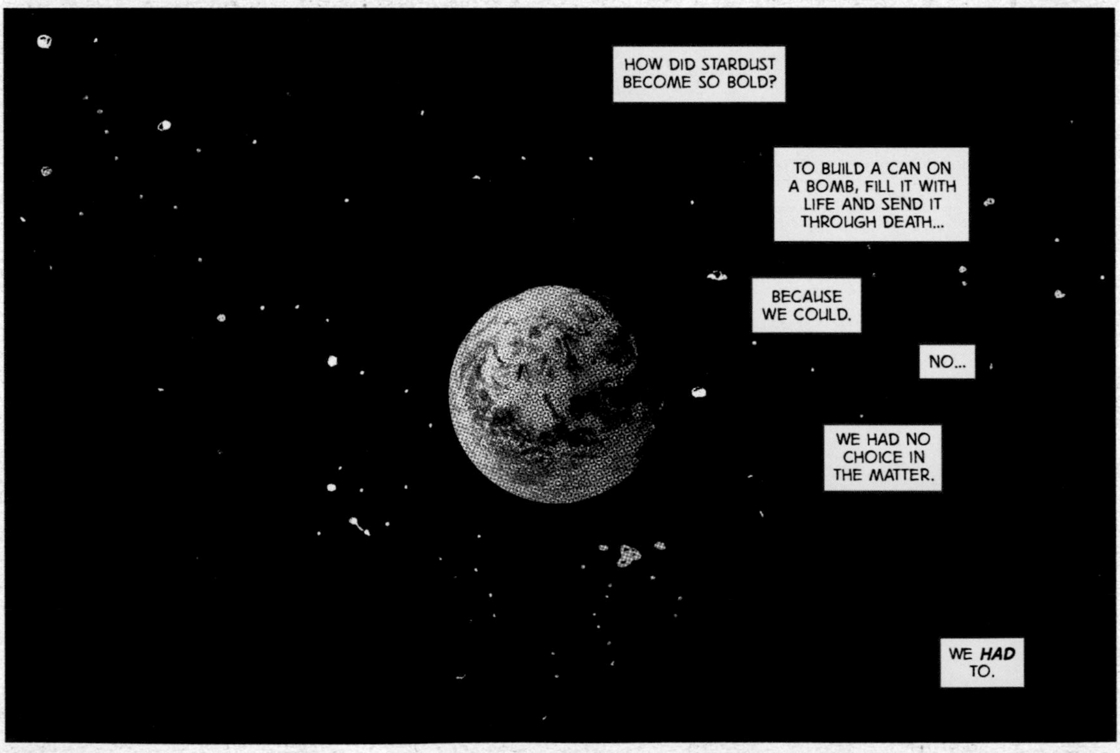
HOW DID STARDUST BECOME SO BOLD?
TO BUILD A CAN ON A BOMB, FILL IT WITH LIFE AND SEND IT THROUGH DEATH...
BECAUSE WE COULD.
NO...
WE HAD NO CHOICE IN THE MATTER.
WE ***HAD*** TO.

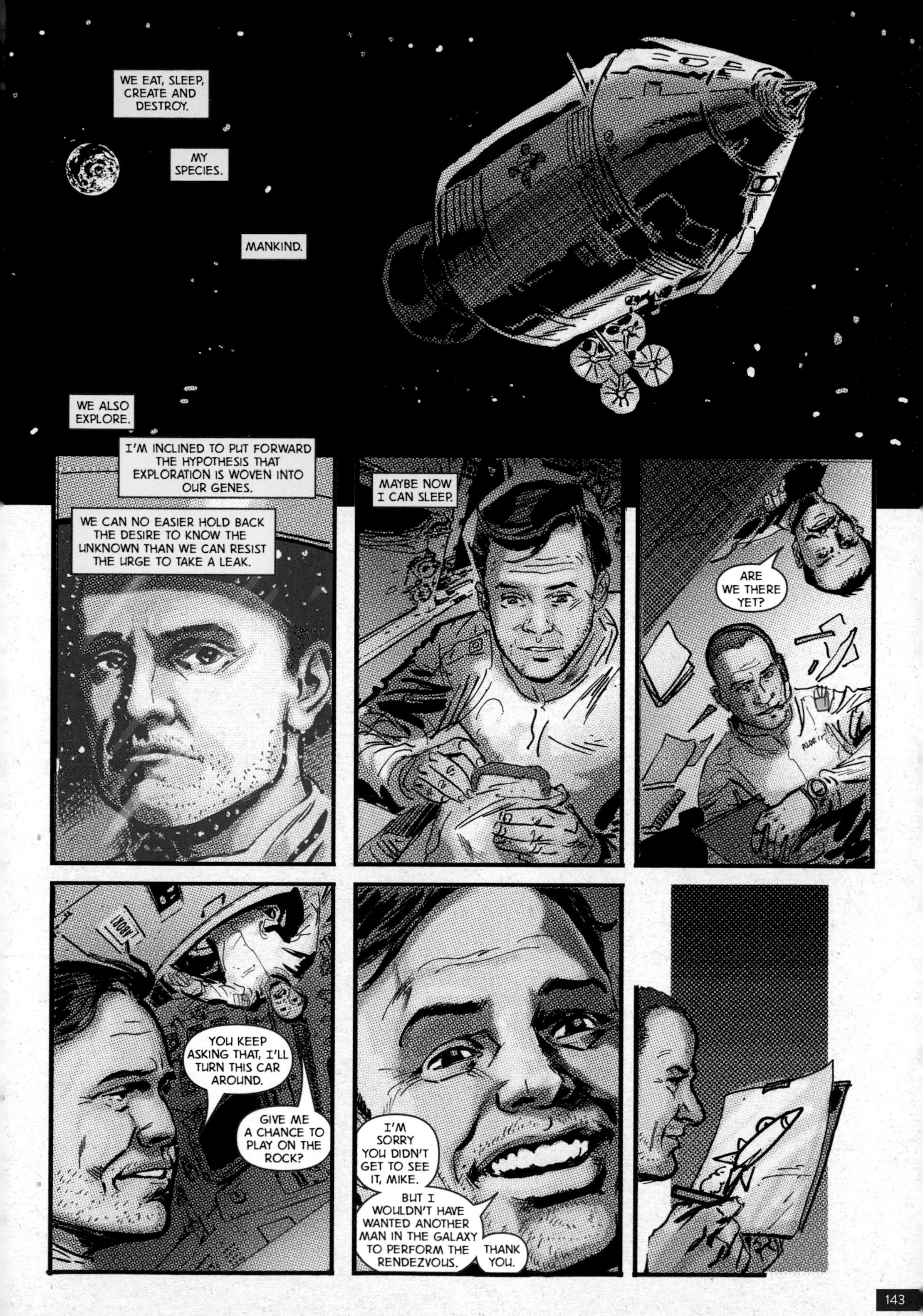
WE EAT, SLEEP, CREATE AND DESTROY.
MY SPECIES.
MANKIND.
WE ALSO EXPLORE.
I'M INCLINED TO PUT FORWARD THE HYPOTHESIS THAT EXPLORATION IS WOVEN INTO OUR GENES.
WE CAN NO EASIER HOLD BACK THE DESIRE TO KNOW THE UNKNOWN THAN WE CAN RESIST THE URGE TO TAKE A LEAK.
MAYBE NOW I CAN SLEEP.
ARE WE THERE YET?
YOU KEEP ASKING THAT, I'LL TURN THIS CAR AROUND.
GIVE ME A CHANCE TO PLAY ON THE ROCK?
I'M SORRY YOU DIDN'T GET TO SEE IT, MIKE.
BUT I WOULDN'T HAVE WANTED ANOTHER MAN IN THE GALAXY TO PERFORM THE RENDEZVOUS.
THANK YOU.

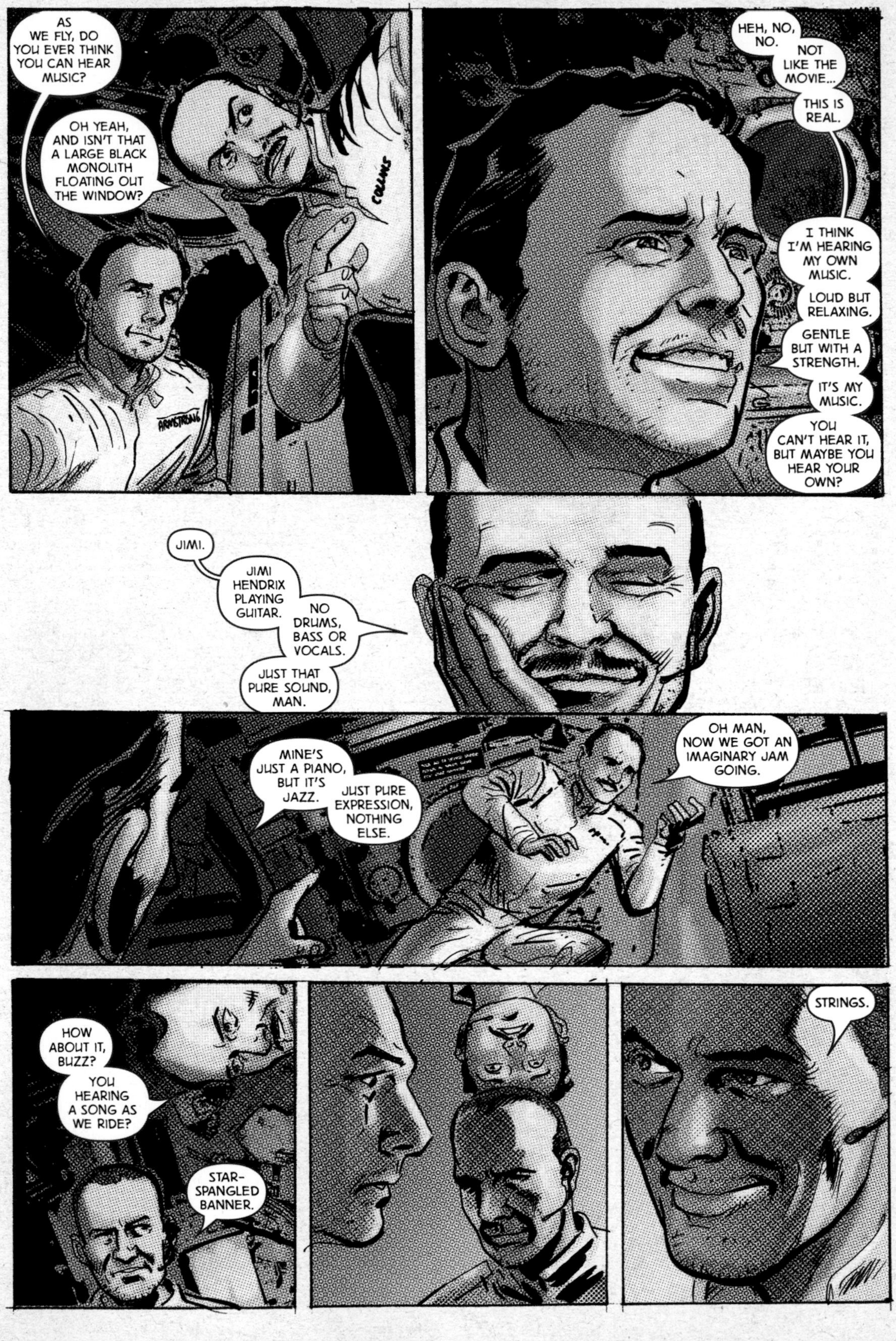
AS WE FLY, DO YOU EVER THINK YOU CAN HEAR MUSIC?
OH YEAH, AND ISN'T THAT A LARGE BLACK MONOLITH FLOATING OUT THE WINDOW?
COLLINS
ARMSTRONG
HEH, NO, NO.
NOT LIKE THE MOVIE...
THIS IS REAL.
I THINK I'M HEARING MY OWN MUSIC.
LOUD BUT RELAXING.
GENTLE BUT WITH A STRENGTH.
IT'S MY MUSIC.
YOU CAN'T HEAR IT, BUT MAYBE YOU HEAR YOUR OWN?
JIMI.
JIMI HENDRIX PLAYING GUITAR.
NO DRUMS, BASS OR VOCALS.
JUST THAT PURE SOUND, MAN.
MINE'S JUST A PIANO, BUT IT'S JAZZ.
JUST PURE EXPRESSION, NOTHING ELSE.
OH MAN, NOW WE GOT AN IMAGINARY JAM GOING.
HOW ABOUT IT, BUZZ?
YOU HEARING A SONG AS WE RIDE?
STAR-SPANGLED BANNER.
STRINGS.

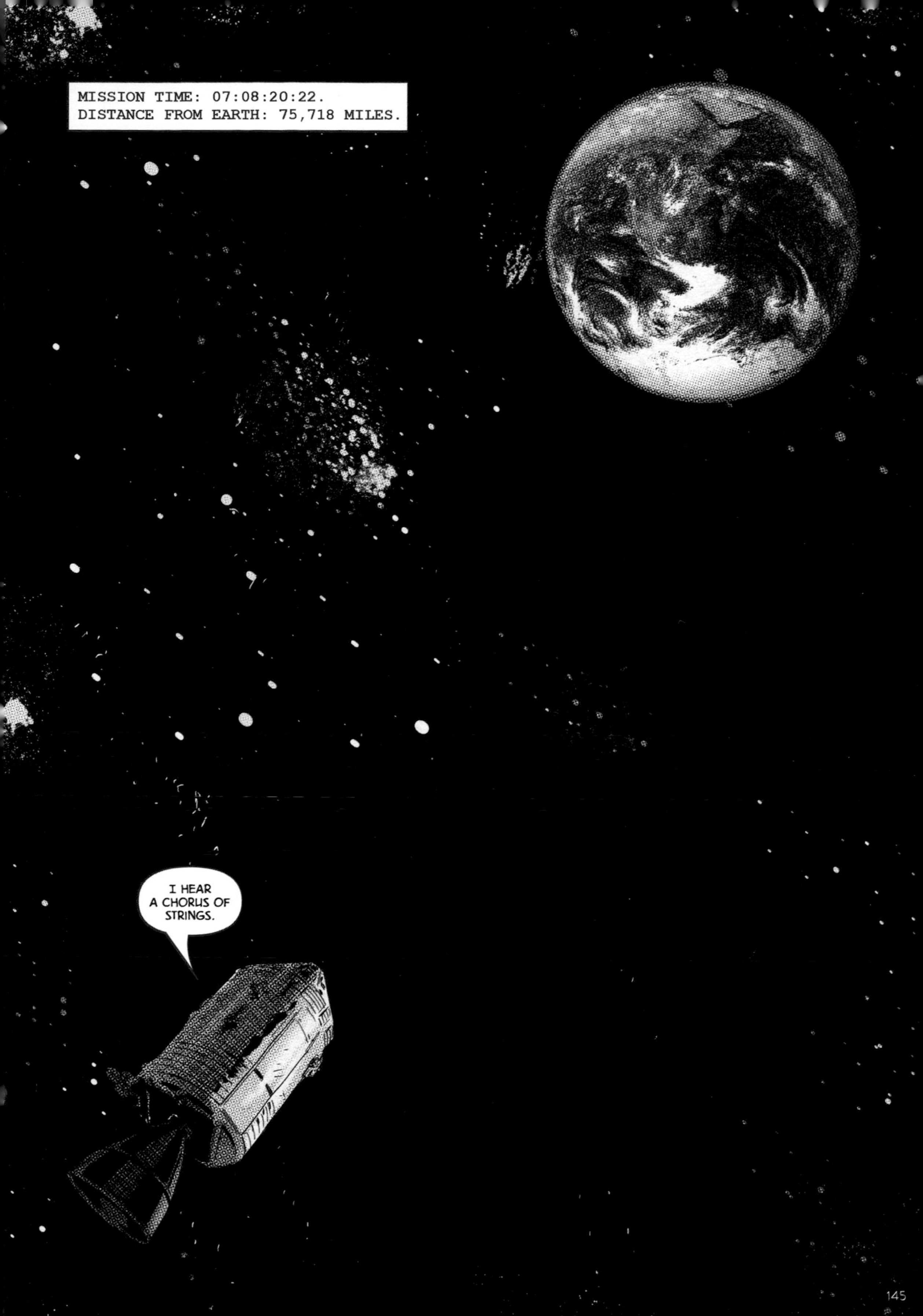
MISSION TIME: 07:08:20:22.
DISTANCE FROM EARTH: 75,718 MILES.
I HEAR A CHORUS OF STRINGS.

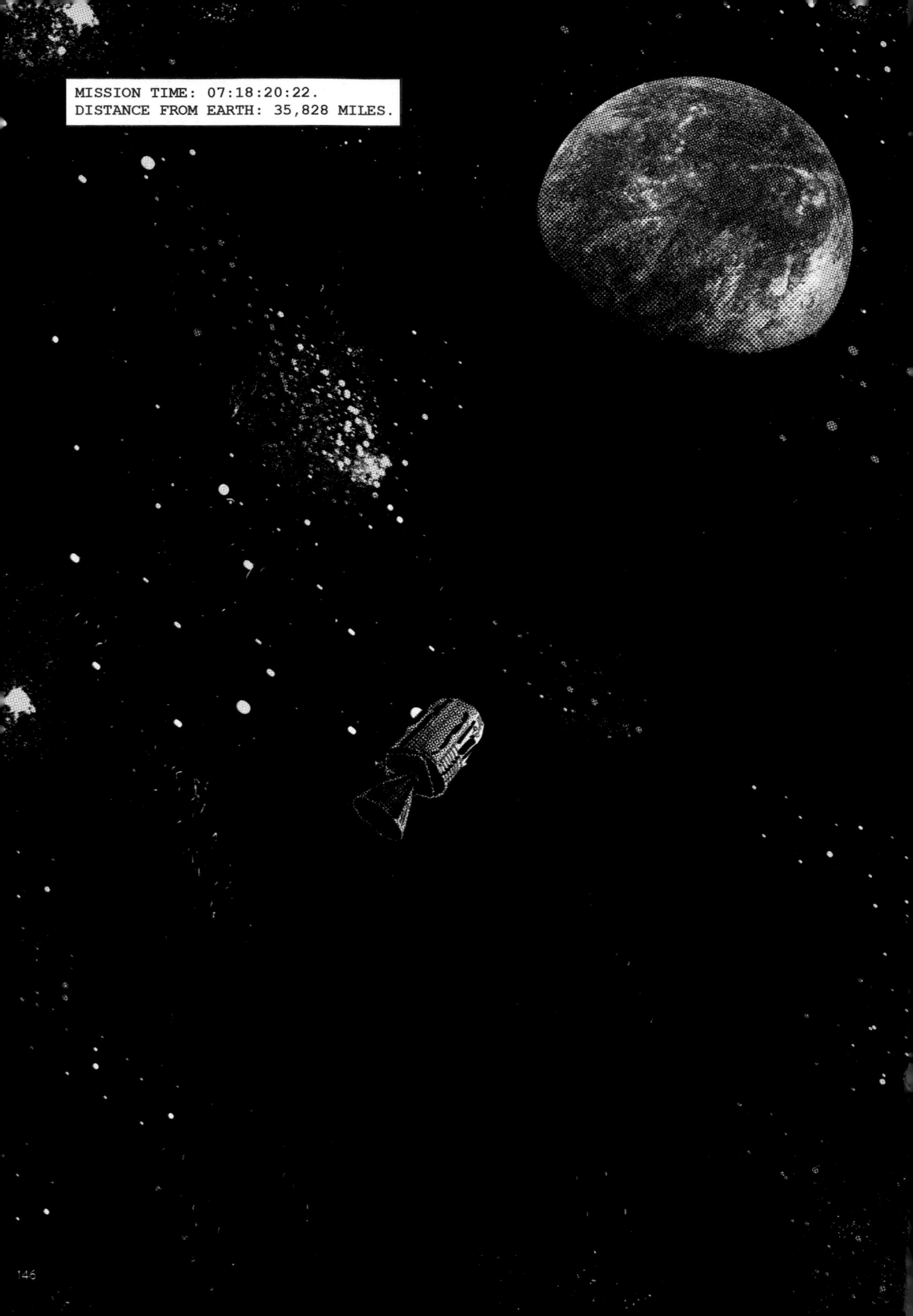
MISSION TIME: 07:18:20:22.
DISTANCE FROM EARTH: 35,828 MILES.

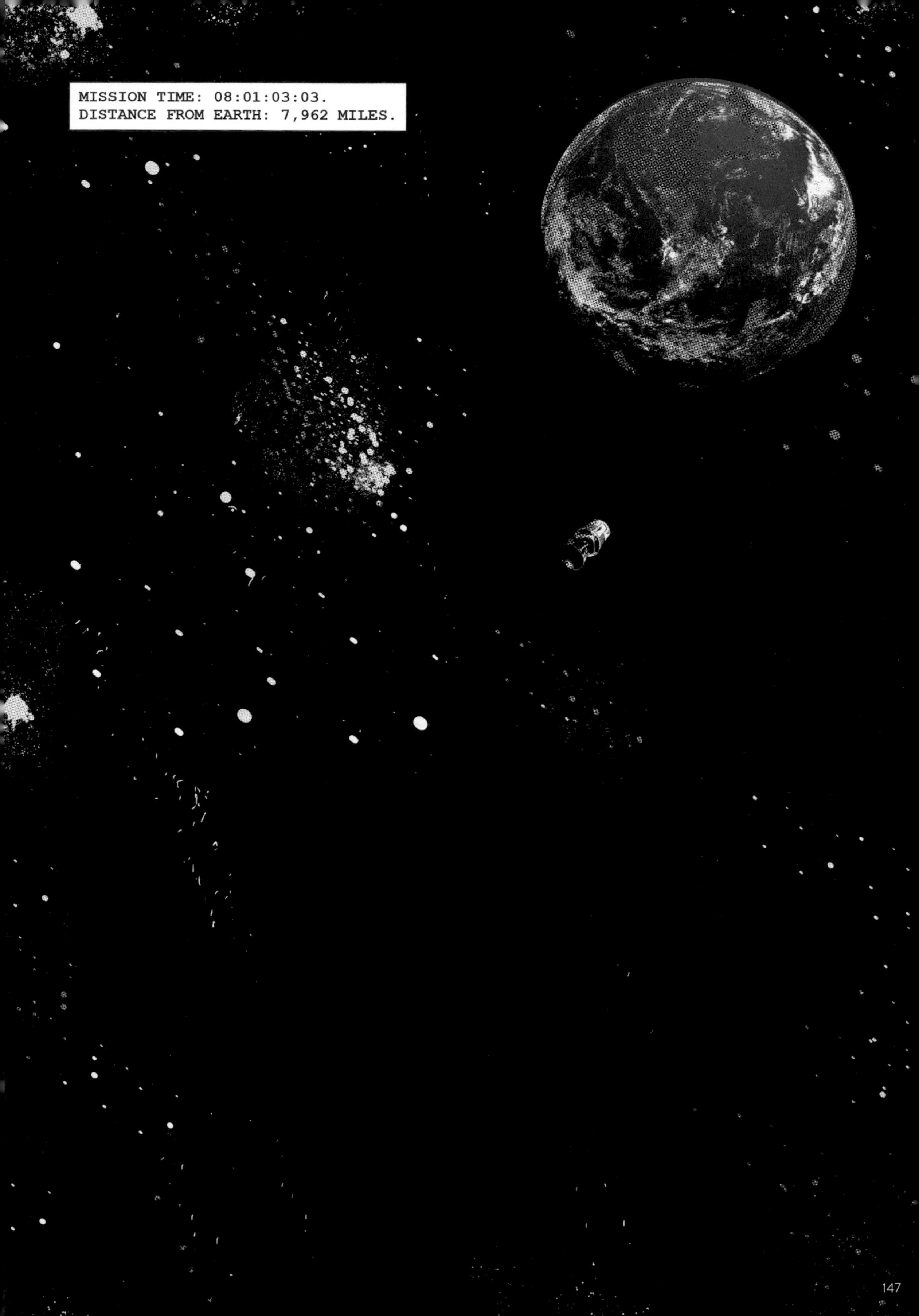
MISSION TIME: 08:01:03:03.
DISTANCE FROM EARTH: 7,962 MILES.

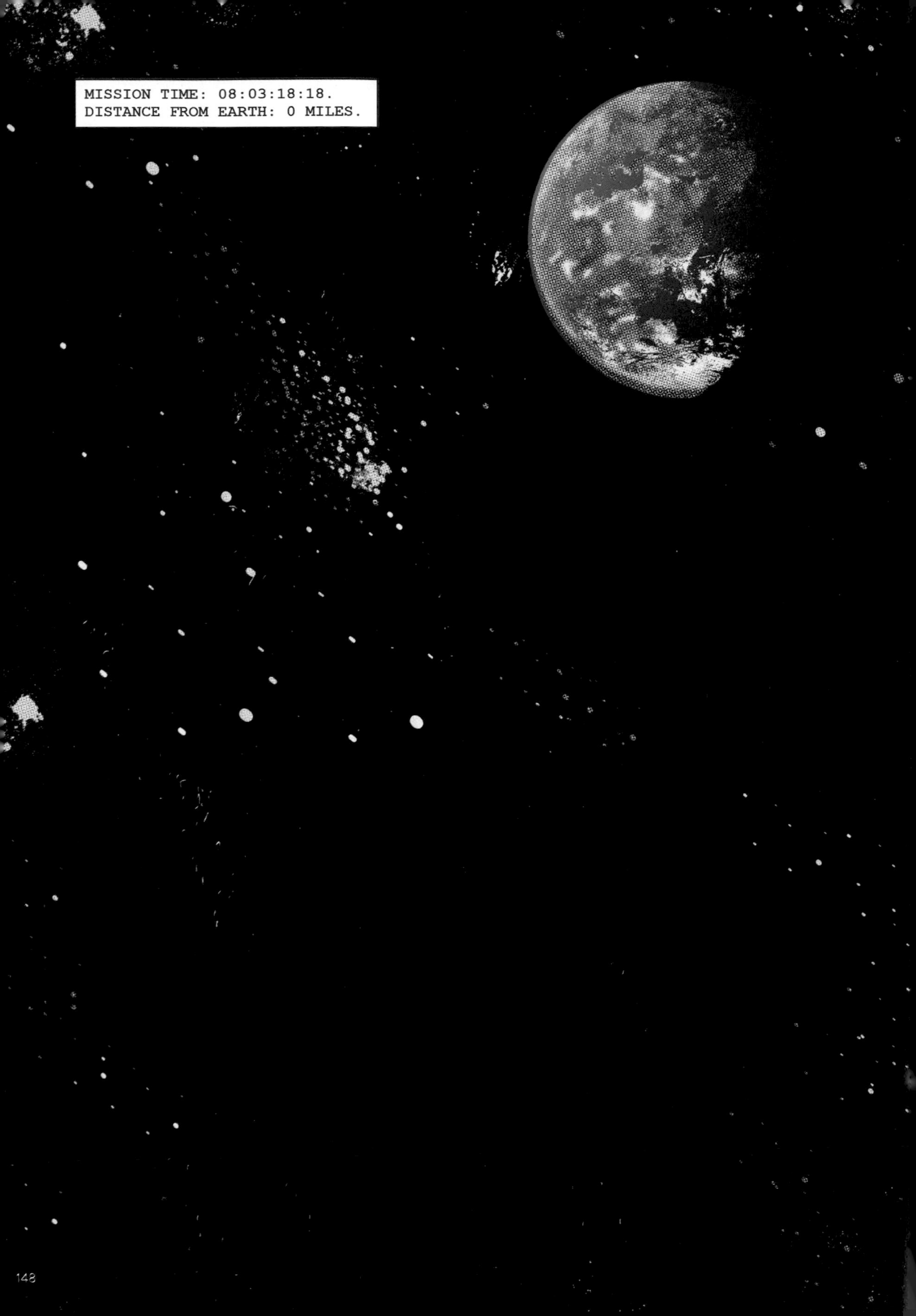
MISSION TIME: 08:03:18:18.
DISTANCE FROM EARTH: 0 MILES.

We set sail on this new sea because there is new knowledge to be gained, and new rights to be won, and they must be won and used for the progress of all people. For space science, like nuclear science and all technology, has no conscience of its own. Whether it will become a force for good or ill depends on man, and only if the United States occupies a position of pre-eminence can we help decide whether this new ocean will be a sea of peace or a new terrifying theatre of war. I do not say that we should or will go unprotected against the hostile misuse of space any more than we go unprotected against the hostile use of land or sea, but I do say that space can be explored and mastered without feeding the fires of war, without repeating the mistakes that man has made in extending his writ around this globe of ours.

There is no strife, no prejudice, no national conflict in outer space as yet. Its hazards are hostile to us all. Its conquest deserves the best of all mankind, and its opportunity for peaceful cooperation may never come again. But why, some say, the Moon? Why choose this as our goal? And they may well ask, why climb the highest mountain? Why, 35 years ago, fly the Atlantic? Why does Rice play Texas?

We choose to go to the Moon! We choose to go to the Moon in this decade and do the other things, not because they are easy, but because they are hard; because that goal will serve to organise and measure the best of our energies and skills, because that challenge is one that we are willing to accept, one we are unwilling to postpone and one we intend to win.

- John F. Kennedy, September 12 1962

APPENDIX

SATURN V

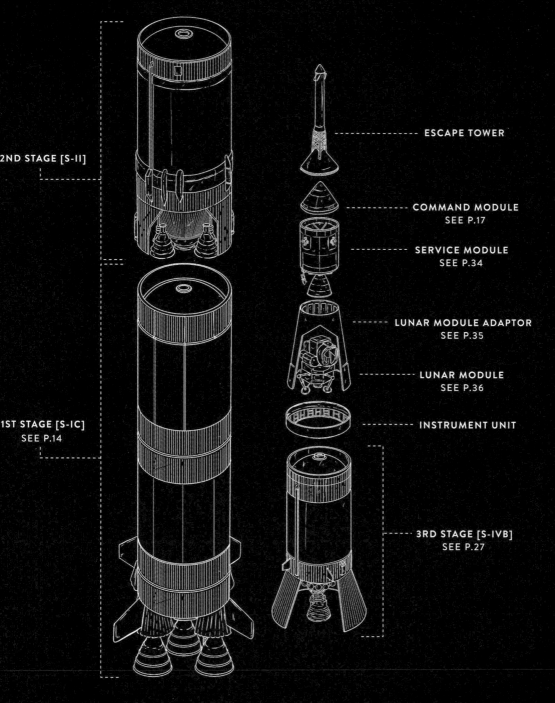

2ND STAGE [S-II]

ESCAPE TOWER

COMMAND MODULE
SEE P.17

SERVICE MODULE
SEE P.34

LUNAR MODULE ADAPTOR
SEE P.35

LUNAR MODULE
SEE P.36

1ST STAGE [S-IC]
SEE P.14

INSTRUMENT UNIT

3RD STAGE [S-IVB]
SEE P.27

THE SATURN V WAS DESIGNED UNDER THE DIRECTION OF WERNHER VON BRAUN AND
ARTHUR RUDOLPH, WITH BOEING, NORTH AMERICAN AVIATION, DOUGLAS
AIRCRAFT COMPANY AND IBM AS THE LEAD CONTRACTORS.

COMMAND/SERVICE MODULE

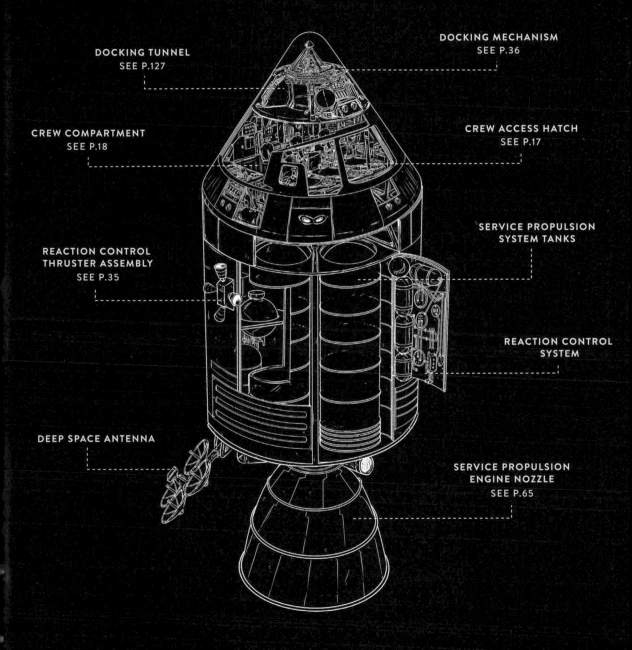

DOCKING TUNNEL
SEE P.127

DOCKING MECHANISM
SEE P.36

CREW COMPARTMENT
SEE P.18

CREW ACCESS HATCH
SEE P.17

REACTION CONTROL
THRUSTER ASSEMBLY
SEE P.35

SERVICE PROPULSION
SYSTEM TANKS

REACTION CONTROL
SYSTEM

DEEP SPACE ANTENNA

SERVICE PROPULSION
ENGINE NOZZLE
SEE P.65

THE COMMAND/SERVICE MODULE [CSM] WAS BUILT FOR THE
APOLLO PROGRAM BY NORTH AMERICAN AVIATION.

LUNAR MODULE

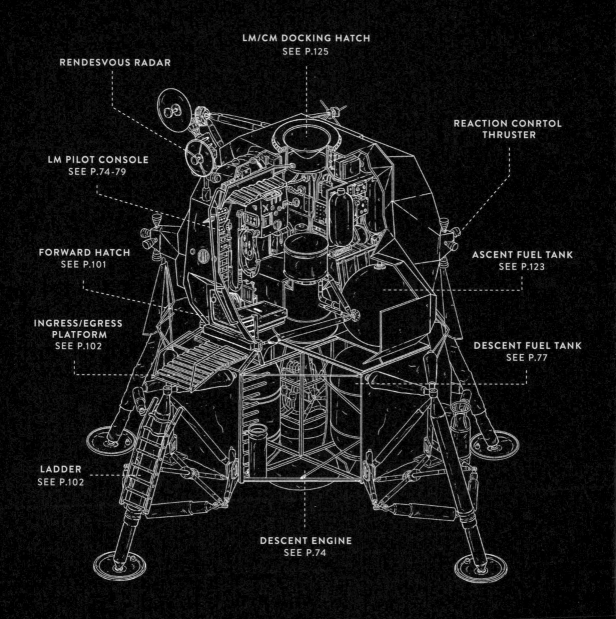

RENDESVOUS RADAR

LM/CM DOCKING HATCH
SEE P.125

REACTION CONRTOL
THRUSTER

LM PILOT CONSOLE
SEE P.74-79

FORWARD HATCH
SEE P.101

ASCENT FUEL TANK
SEE P.123

INGRESS/EGRESS
PLATFORM
SEE P.102

DESCENT FUEL TANK
SEE P.77

LADDER
SEE P.102

DESCENT ENGINE
SEE P.74

THE LUNAR MODULE [LM] WAS BUILT FOR THE APOLLO PROGRAM BY
GRUMMAN AIRCRAFT AND CHIEFLY DESIGNED BY THOMAS J. KELLY

LIST OF NASA ASTRONAUTS FEATURED IN *APOLLO*

PAGE 10:

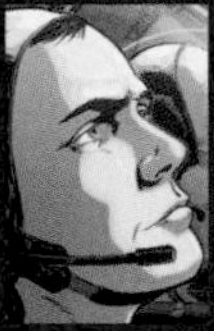

Neil Alden Armstrong: one of the two first humans to land on the Moon; first person to walk on the Moon. Command pilot, Gemini 8 (1966, first successful docking of two vehicles in space). Commander, Apollo 11 (1969, first manned lunar landing).

PAGE 11:

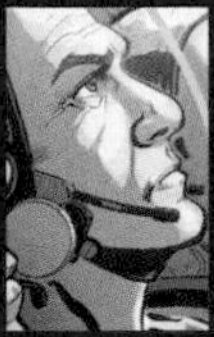

Edwin 'Buzz' Aldrin Jr: one of the two first humans to land on the Moon; second person to walk on the Moon. Pilot, Gemini 12 (1966). LM pilot, Apollo 11 (1969, first manned lunar landing).

PAGE 11:

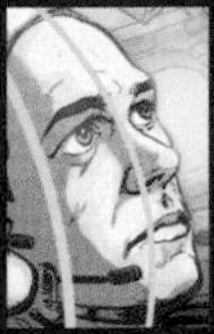

Michael Collins: first person to meet another spacecraft in orbit. Pilot, Gemini 10 (1966). Command module pilot, Apollo 11 (1969, first manned lunar landing).

PAGE 15:

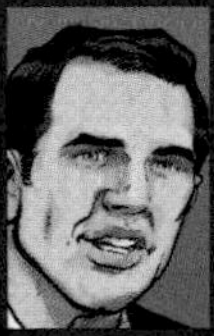

David R. Scott: seventh man to walk on the Moon. Pilot, Gemini 8 (1966, first successful docking of two vehicles in space). Command module pilot, Apollo 9 (1969, third manned flight in Apollo series). Commander, Apollo 15 (1971, third manned lunar landing).

PAGE 17:

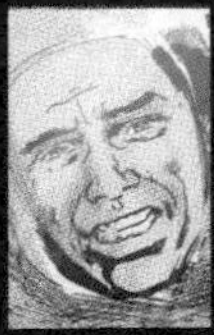

Roger Bruce Chaffee: pilot, Apollo 1 (1967).

PAGE 18:

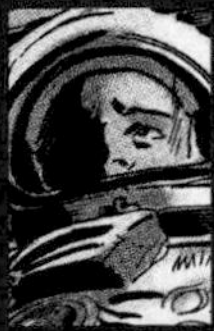

Edward Higgins 'Ed' White: first American to walk in space. Pilot, Gemini 4 (1966, first space walk by an American). Command pilot, Apollo 1 (1967).

PAGE 18:

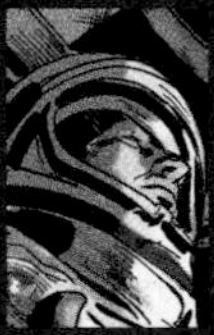

Virgil Ivan 'Gus' Grissolm: second American in space. Pilot, Mercury-Redstone 4 (1961, second American human spaceflight). Pilot, Gemini 3 (1965). Commander, Apollo 1 (1967).

PAGE 21:

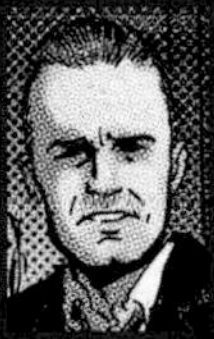

Richard Francis 'Dick' Gordon Jr: pilot, Gemini 11 (1966, first artificial gravity experiment). Command module pilot, Apollo 12 (1969, second manned lunar landing).

PAGE 21:

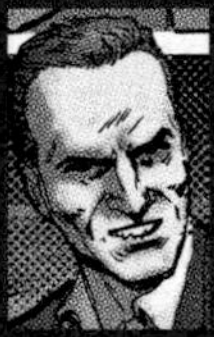

Malcolm Scott Carpenter: second American to orbit the Earth, fourth American in space. Pilot, Mercury-Atlas 7 (1962).

PAGE 22:

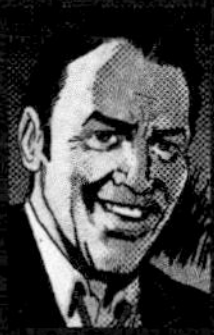

James Arthur 'Jim' Lovell Jr: pilot, Gemini 7 (1965). Commander, Gemini 12 (1966). Command module pilot, Apollo 8 (1968, first manned spacecraft to leave Earth orbit, reach the Moon, orbit it and return safely to Earth). Commander, Apollo 13 (aborted lunar landing, farthest humans have ever travelled from Earth).

PAGE 22:

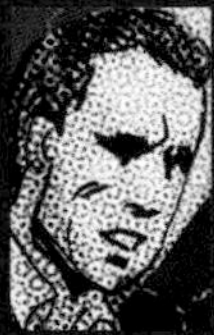

Leroy Gordon 'Gordo' Cooper Jr: first American to sleep in space. Pilot, Mercury-Atlas 9 (1963). Pilot, Gemini 5 (1965).

PAGE 62:

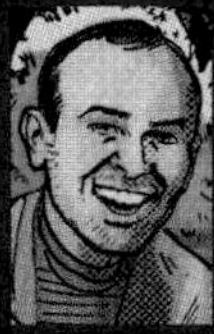

Alan LaVern Bean: fourth person to walk on the Moon. LM pilot, Apollo 12 (1969, second manned lunar landing). Spacecraft commander, Skylab 3 (1973, second manned mission to the first American space station).

BIBLIOGRAPHY & REFERENCES

BOOKS

Aldrin Jr., E., *Magnificent Desolation*, StarBuzz LLC, 2009

Aldrin Jr., E., *Reaching for the Moon*, Harper Collins, 2005

Barbree, J., *Neil Armstrong: A Life of Flight*, Thomas Dunne, 2014

Barbree, J., Slayton, D. and Shepard, A., *Moon Shot: The Inside Story of America's Race to the Moon*, Turner Publishing, 1994

Chaikin, A., *A Man on the Moon: The Voyages of the Apollo Astronauts*, Viking, 1994

Collins, M., *Carrying the Fire: An Astronaut's Journey*, Cooper Square Press, 1974

Gorn, M.H., *NASA: The Complete Illustrated History*, Merrell, 2005

Hansen, J.R., *First Man: The Life of Neil Armstrong*, Simon & Schuster, 2005

Kranz, G., *Failure Is Not an Option*, Simon & Schuster, 2000

Smith, A., *Moon Dust*, Bloomsbury, 2005

Woods, D., *NASA Saturn V Manual*, Haynes Publishing, 2016

OTHER SOURCES

Orloff, R.W., *Apollo by the Numbers: A Statistical Reference*, Library of Congress Cataloging-in-Publication Data, 2000

Apollo 11 Flight Plan, NASA, July 1969 (http://www.hq.nasa.gov/alsj/a11/a11fltpln_final_reformat.pdf)

Apollo 11 Onboard Voice Transcription, NASA, August 1969 (http://uploads.worldlibrary.org/uploads/pdf/20130422214707as11_cm_pdf.pdf)

Johnson Space Center Apollo 11 Mission Transcripts, National Aeronautics and Space Administration, Johnson Space Center, 2010

Apollo 11 Technical Air-to-Ground Voice Transcription, NASA (http://www.hq.nasa.gov/alsj/a11/a11transcript_tec.html)

From the Earth to the Moon (HBO mini-series, 1998)

In the Shadow of the Moon (dir. David Sington and Christopher Riley, 2007)

Apollo 13 (dir. Ron Howard, 1995)

WEBSITES

Spacelog: Apollo 11 (http://apollo11.spacelog.org)

Project Apollo Archive by Apollo Image Gallery on Flickr (http://www.flickr.com/photos/projectapolloarchive/albums)

SPECIAL THANKS

Arts Council England

Science Museum, London

Colin A. Fries, NASA History Office

D. Chris Cottrill, Smithsonian Libraries

Paul Clark-Forse for his valuable editing advice

Orlando Wood

Mark Lewis

Charlie Hodgson

Joanne Caldwell

Helena Delgado-Cohen

ABOUT THE CREATORS

Matt Fitch and **Chris Baker** are a writing team based in the UK. The pair first met while working in the world of advertising and went on to co-found the micro-publisher Dead Canary Comics, where they began to publish their stories and work with some of the most exciting up-and-coming UK artists. Their first graphic novel, *Reddin*, gained noteworthy praise from within the indie comics scene, and since then the two have collaborated on many writing projects. *APOLLO* is their first major publication.

Best known these days as a storyboard artist for TV (Doctor Who, Sherlock, Knightfall, Good Omens), **Mike Collins** has a thirty-year track record as a comic artist and writer, including producing a critically acclaimed adaptation of Dickens' *A Christmas Carol*. He has worked on all the major Marvel and DC characters, including Spider-Man, the X-Men, Superman, Batman and Wonder Woman. He also draws a series of graphic novels based on the adventures of Varg Veum, Gunnar Staalesen's iconic Norwegian private eye.

Ian Sharman is a writer (*Alpha Gods*, *Hero: 9 to 5*, *Hypergirl*, *The Intergalactic Adventures of Zakk Ridley*), editor (the Eagle Award-nominated anthology *Eleventh Hour*), inker (Spider-Man, Iron Man and the X-Men for Panini Comics) and letterer (BBC Books, Top Shelf). He is the editor-in-chief of Markosia and runs Orang Utan Comics, a UK-based collective of comic book creators.

Kris Carter is a colourist and illustrator based in South Wales, UK. He has provided colours for Transformers and Doctor Who, and also for TMNT, Torchwood and Wynona Earp, among others. He illustrated the award-winning *Caring For My Family With Cancer* for Velindre Cancer Centre. Kris is also the co-founder of small-press publisher Attic Studios.

Jason Cardy is a professional artist, colourist and designer who has worked in comics (and related industries) since 2003. He has worked on franchises including Transformers, TMNT and Thundercats, as well as various Marvel properties. He has contributed to several award-winning books, including Classical Comics' *Sweeney Todd* and *Frankenstein*. He lives in South Wales, UK.

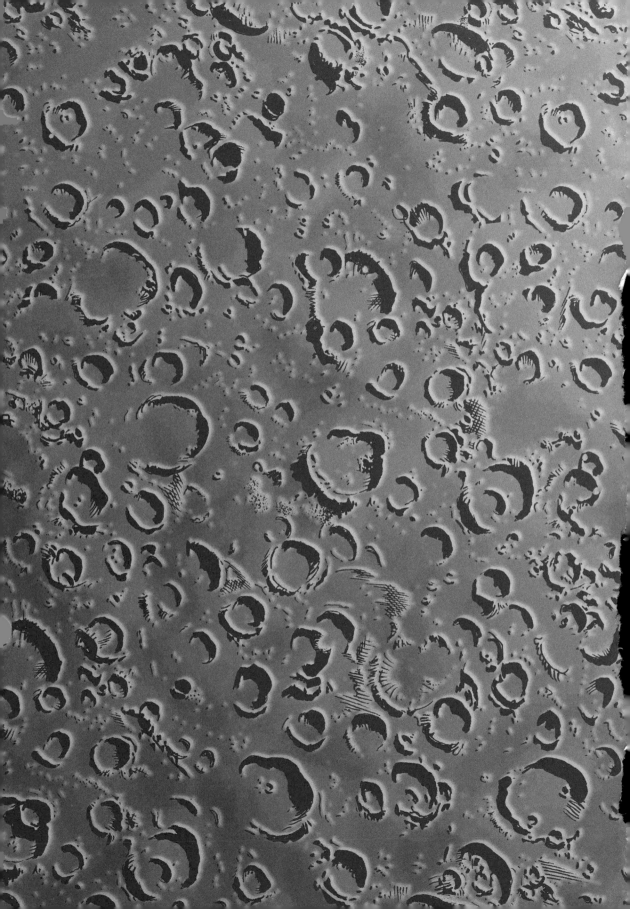